Pollution Abatement Strategies in Central and Eastern Europe

edited by
Michael A. Toman

Resources for the Future
New York

Published 1994 by Resources for the Future
2 Park Square, Milton Park, Abingdon, Oxon, OX14 4RN
711 Third Avenue, New York, NY 10017

Library of Congress Cataloging-in-Publication Data

Pollution abatement strategies in Central and Eastern Europe / edited by Michael A.
    Toman
            p.   cm.
        ISBN 0–915707–73–X (alk. paper) : $19.95
        1. Air pollution—Economic aspects—Central Europe. 2. Air pollution—
    Economic aspects—Europe, Eastern. 3. Environmental policy—Central Europe.
    4. Environmental policy—Europe, Eastern.
    I. Toman, Michael A.
    HC244.Z9A466 1994
    363.73'92—dc20
    [0943]
                                                                94-17714
                                                                CIP

This book is the product of the Energy and Natural Resources Division at Resources for the Future, Douglas R. Bohi, director. It was edited by Melissa K. Edeburn and Betsy Kulamer and designed by Brigitte Coulton. The cover was designed by Kelly Design.

RESOURCES FOR THE FUTURE (RFF) is an independent nonprofit organization engaged in research and public education on natural resource and environmental issues. Its mission is to create and disseminate knowledge that helps people make better decisions about the conservation and use of their natural resources and the environment. RFF neither lobbies nor takes positions on current policy issues.

Because the work of RFF focuses on how people make use of scarce resources, its primary research discipline is economics. However, its staff also includes social scientists from other fields, ecologists, environmental health scientists, meteorologists, and engineers. Staff members pursue a wide variety of interests, including forest economics, recycling, multiple use of public lands, the costs and benefits of pollution control, endangered species, energy and national security, hazardous waste policy, climate resources, and quantitative risk assessment.

Acting on the conviction that good research and policy analysis must be put into service to be truly useful, RFF communicates its findings to government and industry officials, public interest advocacy groups, nonprofit organizations, academic researchers, and the press. It produces a range of publications and sponsors conferences, seminars, workshops, and briefings. Staff members write articles for journals, magazines, and newspapers, provide expert testimony, and serve on public and private advisory committees. The views they express are in all cases their own, and do not represent positions held by RFF, its officers, or trustees.

Established in 1952, RFF derives its operating budget in approximately equal amounts from three sources: investment income from a reserve fund, government grants, and contributions from corporations, foundations, and individuals. (Corporate support cannot be earmarked for specific research projects.) Some 45 percent of RFF's total funding is unrestricted, which provides crucial support for its foundational research and outreach and educational operations. RFF is a publicly funded organization under Section 501(c)(3) of the Internal Revenue Code, and all contributions to its work are tax deductible.

Contents

Foreword

One of the ironies of the former centrally planned economies, we have come to learn, is how little they cared about protecting their environments. Such rapacious behavior should not be so prevalent in societies whose purported objectives were defined in terms of the social rather than the private good. Nevertheless, virtually every one of the countries of Central and Eastern Europe is confronting a Herculean task in slowing down the rate of pollution and cleaning up decades of environmental neglect.

The awareness of the extent of the problem has prompted many Western industrial countries to come to the aid of the Central and Eastern European countries, partly out of concern for the residents of those countries and partly out of concern with the spillover effects of their pollution on the rest of the world. The West has realized, quite practically, that the economic and social problems in Central and Eastern Europe are so vast, and of such higher priority than their environmental objectives, that these countries must have outside help if they are to make any headway on the environmental front in the foreseeable future.

This book is not another litany of horror stories about environmental degradation in Central and Eastern Europe. Rather, it is concerned with the practical issues of what to do about the problems. The book is concerned, first, with assessing the seriousness of the environmental problems in these countries so that priorities can be established for spending scarce resources in the cleanup effort. The priority-setting is especially important because the worst problems from a public health perspective are not always the most obvious ones. This book also examines how high-priority problems can be managed in a realistic and effective way, given existing institutions and scarce resources. In this respect, the book brings to the study of the issues a strong social science perspective, one that is noticeably lacking in Central and Eastern European countries.

As the authors emphasize throughout the book, accomplishing real progress in cleaning up the environment in these countries requires attention to both cost-effectiveness and the capabilities of institutions in transition. Expensive and draconian approaches, like existing environmental regulations in these countries, will simply be ignored. Approaches that work well in Western industrial countries may be entirely unrealistic in these countries because their success depends upon well-established economic and legal institutions. In this respect, the book makes an important contribution to our understanding of the issues involved in environmental management in a transition economy.

The essays in *Pollution Abatement Strategies in Central and Eastern Europe* are not intended only for experts in the environmental field. Public and private decision makers will find many insights that can help improve their understanding of the environmental problems of Central and Eastern Europe, as well as practical suggestions for overcoming them. Students in economics, public policy, international affairs, and environmental studies will find this book an excellent demonstration of how applied economics can be used to address important environmental and social problems.

Douglas R. Bohi
Senior Fellow and Director
Energy and Natural Resources Division
Resources for the Future

Preface

Starting in the latter half of 1990, as information accumulated about environmental problems in the newly emerging countries of Central and Eastern Europe, a number of researchers from Resources for the Future (RFF) began to discuss what research might be useful to help address these problems. We were fortunate, therefore, when RFF was approached in early 1992 by representatives of the U.S. Environmental Protection Agency (EPA) and the World Bank to determine our interest in undertaking a fairly large-scale analysis of some of these problems. The World Bank faced the challenging task of putting together an Environmental Action Programme for Central and Eastern Europe, a blueprint for environment and finance ministers concerning how that region and countries in the West might best respond to the environmental challenges in the region. EPA was interested in assisting the process and thought that RFF research could contribute to the underpinnings of the Programme. In the spring of 1992, following agreement on the tasks to be undertaken and the funding from EPA, the RFF team began work. Subsequent additions to the work plan reflected growing interest by the U.S. Agency for International Development, the World Bank, and RFF researchers themselves in the problems facing the former Soviet Union.

This volume represents a distillation of RFF's efforts. Seven of the chapters first appeared in *Resources*, RFF's quarterly periodical, and the last chapter is entirely new. All the chapters are drawn from more extensive studies; readers who wish to obtain more detailed discussions about methodology and results should contact the authors. None of the chapters in this book has received formal peer review, but the authors have benefited from considerable amounts of less formal comment on the results of their research.

From the start, our prior experience in analyzing environmental problems led us to emphasize standard themes in environmental economics: How bad are the problems at hand? What are the options for ameliorating them in an economically effective fashion? As time went on, however, two other themes emerged that may be at least as significant for environmental management in Central and Eastern Europe: What can be done in practice, given the institutional constraints of a transitional economic and political system? In particular, how are environmental problems and solutions bound up with economic problems and their solutions?

Each of the chapters in this book addresses one or more of these basic themes. In the first chapter, Margaret A. Walls examines potential changes in future levels of motor vehicle emissions in four Central and Eastern European countries. These changes will reflect both increases in motor vehicle use, which will increase emissions, and improvements in cars' environmental performance, which will reduce emissions. In the second chapter, Alan J. Krupnick, Kenneth W. Harrison, Eric J. Nickell, and I estimate the value of human health benefits that would be realized through air quality improvements in five Central and Eastern European countries.

In the third chapter, I examine the potential changes over time in air pollution that result from energy use in Poland. I also consider the advantages offered by incentive-based environmental policies—such as emission permit trading—for reducing this pollution and compare incentive-based policies with command-and-control environmental policies. The fourth chapter contains a case study of the Nitra River basin in the Slovak Republic, which Charles M. Paulsen uses to compare the cost-effectiveness of incentive-based policies and command-and-control policies for controlling point sources of pollutant discharges into rivers and lakes in Central and Eastern Europe.

The fifth chapter, by Dallas Burtraw, analyzes the obstacles to and the benefits of sulfur dioxide emission permit trading among electric utilities in both Western Europe and Central and Eastern Europe. In the sixth chapter, James Boyd assesses various systems of assigning liability for existing soil and groundwater contamination in the region. Specifically, he considers several policy options for allocating pollution cleanup costs between governments and current or future property owners.

The last two chapters consider institutional issues that go beyond the design of environmental policies. R. David Simpson and I examine how broader institutional failures related to environmental management hamper the implementation of *any* environmental policies, especially in the former Soviet Union. Michael T. Rock examines economic development policies that may promote environmental improvement and may be among the lowest-cost measures for environmental management.

Several common conclusions emerge from these investigations. One is that some air pollution problems in Central and Eastern Europe may not be as serious or as ubiquitous as once thought. For example, Walls finds that most air pollution problems due to motor vehicle use are not any more severe in the countries she studied than they are in the United States. Moreover, she finds that these problems need not worsen significantly by 2010, although she cautions that ozone problems caused by motor vehicle emissions of hydrocarbons and nitrogen oxides may be of concern in urban areas. My discussion of the Poland case study indicates that air pollution could be substantially reduced as a consequence of energy conservation brought on by both higher energy prices and economic rationalization, quite apart from explicit environmental policies.

A second conclusion is that, given the need to invest heavily in economic restructuring, Central and Eastern European countries must ensure that resources used in pollution control efforts will be directed to the efforts that yield the greatest benefits to society. Walls' recommendations concerning control of motor vehicle emissions suggest that Central and Eastern Europe should focus on pollution control strategies that target the worst pollution problems, rather than on strategies that address pollution problems of relatively little concern and therefore produce few benefits. Because the health benefits of reductions in emissions of particulates and lead could be quite large, Krupnick, Harrison, Nickell, and I assert that air pollution control should be a target of environmental and economic policies in Central and Eastern Europe. And because a U.S.-style system of retroactive environmental liability would encourage the cleanup of only relatively unpolluted properties in the region,

Boyd argues for the establishment of publicly financed liability funds, which could put a priority on pollution mitigation measures at sites presenting the greatest social risks.

A third conclusion is that, given limited resources, Central and Eastern European countries must identify the most cost-effective mechanisms for dealing with their environmental problems. Several of the authors in this book find that incentive-based environmental policy instruments are potentially more cost-effective than command-and-control ones. For example, in the case study of Poland, I find that air pollution emissions can be controlled at less cost through the use of emissions fees and emission permit trading than they can be through the use of certain command-and-control emissions regulations.

Emission permit trading has the further advantage of not creating a large, new tax burden. However, such trading faces several obstacles that could diminish its cost-saving potential and even discourage its implementation. According to Burtraw, achieving the full cost-saving potential of sulfur dioxide emission permit trading in Europe's electricity industry is hampered by the structural and regulatory diversity of that industry. The cost savings from incentive-based policies also may be limited if the command-and-control regime already offers a measure of flexibility. In addition, the opportunity for using emission permit trading may hinge on the scope of the trading. I argue that emission permit trading by individual nations will be beneficial. However, Burtraw asserts that countries are unlikely to adopt such trading, as doing so would mean that their firms would raise product prices in order to reflect the price of emissions permits. As a result, the firms would experience a reduction in their international competitiveness. Burtraw insists that, unless international environmental agreements specify emission permit trading as the mechanism for achieving emissions reduction goals, countries have an important disincentive to engage in such trading.

Even though there are difficulties in implementing incentive-based environmental policies, such policies are worth pursuing in Central and Eastern Europe where possible. For instance, Burtraw asserts that emission permit trading in Europe's electricity industry will promote the adoption of electricity prices that reflect the costs to society of pollution resulting from electricity generation. Moreover, Paulsen notes that command-and-control environmental policies may be too costly for the governments of Central and Eastern Europe to implement. He concludes that the region's severe resource constraints and current institutional fluidity make cost-effective pollution control policies attractive and potentially easier to implement than in the West, where such policies are not the norm.

One key question remains: To what extent are incentive-based policies possible in a transitional setting? I endorse an evolutionary approach to the trading of air pollution emissions in Poland as a way to make pollution control more affordable, thereby promoting environmental improvement. However, this argument explicitly assumes sufficient progress in basic economic, legal, and political restructuring to make a trading regime feasible. As Simpson and I point out, this assumption is especially tenuous at present in Russia, Ukraine, and other countries of the former Soviet Union. (This argument applies also to other Central and Eastern European countries where general reforms are less advanced.) Since basic restructuring is a prerequisite to success in environmental policy and may also be a low-cost strategy for environmental improvement, as Rock observes, support for restructuring efforts should be a priority within Central and Eastern European countries. Moreover, such support should be a priority in the allocation of foreign assistance by Western countries.

In addition to the many individuals who contributed to the preparation of the individual essays in this volume, I would like to thank several people who helped us in the project as a whole. Dan Beardsley at EPA and Richard Ackermann at the World Bank deserve our warm appreciation for their roles in getting the project underway. I am also grateful to RFF's publications staff for its efforts in producing this book. Of the many colleagues and friends in Central and Eastern Europe we have gained over the past two years, I would particularly like to thank Tomasz Zylicz at Warsaw University and Juriy Ruban in the Ukrainian Ministry of Environment for sharing both their wise counsel and their passion for progress in the region.

The new generation of economists and decision makers in Central and Eastern Europe will face a difficult but exciting challenge in forging viable societies and sustainable environments. This book is dedicated to them.

Michael A. Toman
Senior Fellow
Resources for the Future

Motor Vehicles and Pollution in Central and Eastern Europe

Margaret A. Walls

Compared with motor vehicles in the United States, motor vehicles in Central and Eastern Europe are much more polluting, but they are also fewer in number and less used. As a result, both total and per capita motor vehicle emissions of carbon monoxide (CO), hydrocarbons (HC), and nitrogen oxides (NO_x) are lower in Central and Eastern Europe than they are in the United States. Estimates of motor vehicle emissions levels in several Central and Eastern European countries in the near future indicate that these levels may not change substantially relative to population. The estimates, which are based on forecasts of the number of cars and the number of miles driven annually in the countries through the year 2010, suggest that per capita emissions of CO, HC, and NO_x will remain below those in the United States even under a high emissions scenario. They imply that Central and Eastern European countries should avoid costly national regulation of motor vehicle emissions and instead focus on reducing driving in cities, where motor vehicle use poses the most severe air pollution problems, and decreasing the lead content of gasoline.

Air pollution problems in Central and Eastern Europe are well-documented. The problems receiving the most attention are those associated with the use of coal in power plants and heavy industry. Less understood are the air pollution problems that arise from the use of motor vehicles.

The primary pollutants emitted by motor vehicles that run on gasoline are carbon monoxide (CO), hydrocarbons (HC), and nitro-

1

gen oxides (NO_x). Hydrocarbons and nitrogen oxides combine in the atmosphere to form ground-level ozone, the principal ingredient in urban smog. Motor vehicles that run on gasoline containing lead also emit lead into the atmosphere. Those that run on diesel fuel emit particulates and sulfur dioxide (SO_2).

The only accurate way to assess the extent of the pollution problems caused by the use of motor vehicles is to monitor ambient air quality. In the case of problems resulting from the use of leaded gasoline, it is also necessary to analyze the level of lead in the bloodstreams of individuals who are exposed to lead emissions. Unfortunately, in most Central and Eastern European countries there is no extensive testing of individuals' blood-lead levels, and there are few facilities that monitor ambient CO and ozone concentrations.

Given the lack of data on ambient air quality, an alternative way of assessing the extent of air pollution problems associated with motor vehicles in Central and Eastern Europe is to examine patterns of vehicle ownership, use, and emissions in the region. I analyzed such patterns in Bulgaria, the Czech Republic and Slovakia, Hungary, and Poland in order to forecast future levels of emissions from motor vehicles in these countries. (At the time of the analysis, the Czech Republic and Slovakia were one country and therefore are referred to as one country throughout the findings reported below.) Our analysis focused on emissions of CO, HC, NO_x, and lead rather than on particulates and SO_2, as motor vehicles account for less than 5 percent of total emissions of these two pollutants in most Central and Eastern European countries.

Patterns of vehicle ownership and use

Several statistics concerning car ownership and use seem to suggest that pollution problems stemming from motor vehicle use are less severe in Bulgaria, the Czech Republic and Slovakia, Hungary, and Poland than in the countries of the West. First, in these four countries the average number of cars per 1,000 people is 142—about 40 percent of the average ratio of cars to people in the western region of Europe and 25 percent of that in the United States. Second, cars in the four Central and Eastern European countries are driven less than cars in the West. On average, a car in one of these four countries is driven only half as many miles per year as a car in the United States.

Despite the fact that they are fewer in number and are driven fewer miles, cars in Central and Eastern Europe tend to be more polluting than cars in the West. This tendency is connected in part with the fact that cars in Central and Eastern Europe are much older than cars in the West. The average age of cars is 15 years in Bulgaria, the Czech Republic and Slovakia, Hungary, and Poland, but only 7.6 years in the United States. In addition, the percentage of very old cars being driven in these four countries is higher than that in the United States. In the Czech Republic and Slovakia, for example, 48 percent of the cars on the road are more than 10 years old, and in Hungary 42 percent are more than 10 years old. In the United States, however, only 30 percent are this old. Age can be an important factor in how much cars pollute because older cars generally lack the modern pollution control equipment, such as catalytic converters and electronic fuel injection, that newer cars have. In Central and Eastern Europe this problem is compounded by the fact that the region's technology for producing vehicles with such equipment lags behind that of the West.

In addition to their comparatively old age, there are other reasons why cars in Central and Eastern Europe are some of the most polluting vehicles in the world. One is that they are poorly maintained. Another is that they are often defective when they come off the assembly line. Yet another is that they run on highly polluting fuels. The sulfur content of diesel fuel sold in Central and Eastern Europe is high, leading to high levels of particulate emissions. Perhaps more important, nearly all the gasoline sold in the region is leaded gasoline. Moreover, the lead content of that gasoline is typically higher than that of leaded gasoline in Western Europe. While the gasoline in most Western European countries contains 0.15 grams per liter (g/l) of lead, the lead content of gasoline averages between 0.3 g/l and 0.6 g/l in Poland, 0.3 g/l in Hungary, and 0.2 g/l in the Czech Republic and Slovakia.

Emissions from one segment of the vehicle population in Central and Eastern Europe are of particular concern. In most countries of the region, vehicles with two-stroke engines are still being driven. These engines, which burn a mixture of gasoline and lubricating oil, emit a high level of hydrocarbons. Most Central and Eastern European countries have ceased production and banned imports of vehicles with two-stroke engines; however, 9 percent of vehicles driven in Poland, 15 percent of vehicles driven in Bulgaria, and 40 percent of vehicles driven in Hungary have these engines.

Comparison of national motor vehicle emissions

One way to gauge the severity of air pollution problems caused by motor vehicles in Bulgaria, the Czech Republic and Slovakia, Hungary, and Poland is to compare motor vehicle emissions in these countries with those in the United States. To do this, we estimated total annual emissions of CO, HC, NO_x, and lead from motor vehicles in each of the four Central and Eastern European countries in 1990 and in the United States in 1989 and in 1970, when motor vehicle emissions in that country were substantially uncontrolled. In order to account for the differences in the size of each of these countries, we made comparisons of the emissions on the basis of each country's total population.

It is widely held that the United States had unacceptably high ambient CO concentrations in 1970. Until approximately 1980, the national ambient CO standard of 9 parts per million was exceeded in many locations. Although ambient CO concentrations remain unacceptably high during the winter months in some high-altitude cities, the national average CO concentration today is well below the standard. A comparison of per capita CO emissions from motor vehicles in the United States in 1989—132 metric tons per 1,000 people—with those in the four Central and Eastern European countries in 1990— 62 metric tons per 1,000 people—suggests that the latter countries have a less severe CO emissions problem than the United States has even today. Since CO emissions from motor vehicles do not pose much of a problem in the United States, it is possible that they do not pose much of a problem in Bulgaria, the Czech Republic and Slovakia, Hungary, and Poland.

Like CO emissions, lead emissions from motor vehicles were a serious problem in the United States in 1970. Since the country's phaseout of leaded gasoline, such emissions have become virtually a concern of the past. A comparison of per capita lead emissions from motor vehicles in the United States and in the four Central and Eastern European countries suggests that the latter countries have a less severe lead emissions problem than the United States did in 1970, but a more severe problem than the United States did in 1989. In the United States, emissions of lead per 1,000 people dropped from .761 metric tons in 1970 to .008 metric tons in 1989. In the four Central and Eastern European countries, emissions of lead per 1,000 people totalled .045 metric tons. Given their use of leaded gasoline, these countries may have a significant lead problem; however, it is difficult to determine how severe the problem is because data on the blood-lead levels of their city-dwelling citizens are limited.

Compared with CO and lead, ozone has posed a more difficult problem for the United States. Many urban areas continue to violate the U.S. ozone standard. Average ambient ozone concentrations have declined since the early 1970s, but not nearly as much as average ambient CO concentrations have. Emissions of hydrocarbons and nitrogen oxides, the two precursors of ozone, dropped from 46 metric tons and 31 metric tons per 1,000 people in 1970 to 44 metric tons and 24 metric tons per 1,000 people, respectively. While per capita HC and NO_x emissions from motor vehicles in the four Central and Eastern European countries—10 metric tons and 6.5 metric tons per 1,000 people, respectively—are less than those in the United States, it is impossible to conclude from this fact that the former countries do not have an ozone problem.

Comparison of car emissions in cities

One major drawback to the above comparisons of per capita motor vehicle emissions is that, because they are made on a nationwide basis, they do not reflect the fact that air quality problems—particularly those associated with motor vehicles—are inherently problems of urban areas. Therefore we estimated emissions on a city-by-city basis. In doing so, we compared total amounts of HC and CO emissions from cars (rather than emissions from all motor vehicles, as in the above comparisons) in three Central and Eastern European cities—Prague, Sofia, and Budapest—with those in one U.S. city—Milwaukee, Wisconsin.

Because car emissions figures for Central and Eastern European cities are not available, we had to estimate them for Prague, Sofia, and Budapest. We did so on the basis of a calculation involving the number of cars in each of the cities, the number of miles traveled annually per car in each of the countries of which the cities are capitals, and an estimate of the grams-per-mile emissions of U.S. cars prior to 1970. We performed the same calculation to estimate car emissions in Milwaukee, but in place of the estimate of the grams-per-mile emissions of U.S. cars prior to 1970 we used an estimate of the average grams-per-mile emissions of the U.S. car fleet in 1990. We obtained the latter estimate by running the U.S. Environmental Protection Agency's MOBILE 5.0 emissions model.

We chose to compare the car emissions in Milwaukee with those in the three Central and Eastern European cities for several reasons. First, the population of Milwaukee is almost the same as that of Prague and Sofia. Second, the summer temperatures in Milwaukee are very similar to those in Sofia and Budapest. This similarity is important because temperature is a strong predictor of evaporative HC emissions and ambient ozone concentrations. Third, Milwaukee has violated the U.S. ozone standard in recent years. Thus if our HC emissions estimates for Sofia, Prague, and Budapest are equal to or greater than those of Milwaukee, we venture that the former cities might have ozone problems.

According to these estimates, total HC emissions from cars in Budapest are about 30 percent greater than HC emissions from cars in Milwaukee; HC emissions from cars in Prague are about 5 percent greater than those in Milwaukee; and HC emissions from cars in Sofia are about 10 percent less than those in Milwaukee. Thus Prague and Sofia have approximately the same amount of total HC emissions from cars as Milwaukee, and Budapest has a significantly greater amount than Milwaukee, even though the number of cars and the number of miles traveled per car are both much greater in Milwaukee than in the other three cities. There are approximately twice as many cars per 1,000 people in Milwaukee as there are in the three Central and Eastern European cities (560 compared with 340 in Prague, 269 in Sofia, and 250 in Budapest), and cars are driven about twice as many miles per year in Milwaukee as they are in the other three cities.

Since Milwaukee exceeds the U.S. ambient ozone standard, the above estimates suggest that Prague, Sofia, and Budapest may exceed that standard as well. The likelihood that Sofia and Budapest exceed the standard is increased by the fact that they have summer temperatures similar to those of Milwaukee. The likelihood that Prague exceeds the standard is increased by the fact that it has a lower percentage of total HC emissions attributable to motor vehicles than does Milwaukee: 33 percent compared with 42 percent. (Percentages of total HC emissions attributable to motor vehicles in Sofia and in Budapest are unavailable.) This means that Prague must have a greater amount of total HC emissions than Milwaukee and thus is likely to have an ozone problem.

It is more difficult to determine whether the three Central and Eastern European cities have a CO problem. Each has substantially more CO emissions from cars than Milwaukee; but since Milwaukee does not violate the U.S. ambient CO standard, it is difficult to draw any conclusions about the severity of ambient CO conditions in Prague, Sofia, or Budapest.

Forecasts of increases in motor vehicle ownership and use

One of the most important questions for policymakers in Central and Eastern European countries is whether they will have a motor vehicle pollution problem to deal with in the future. Will ownership and use of motor vehicles increase, and, if so, by how much? How will such an increase affect total emissions of HC, CO, NO_x, and lead? Should controls on motor vehicle emissions be required, and, if so, how stringent should they be?

As a starting point in our attempt to answer these questions, we forecasted increases in the number of cars on the road and the number of miles traveled annually per car in Bulgaria, the Czech Republic and Slovakia, Hungary, and Poland through the year 2010. Using data from developed western countries, we established relationships among gasoline prices, gross national product (GNP), and car ownership and use. In doing so, we implicitly assumed that these relationships will hold in the future for Central and Eastern European countries. We then used World Bank GNP forecasts for the four Central and Eastern European countries in question and our best estimates of future gasoline prices to forecast the number of cars on the road and the number of miles traveled annually per car through the year 2010. We then assumed that percentage increases in the number of trucks and motorcycles on the road and percentage increases in the number of miles traveled annually per truck and per motorcycle were the same as those for cars.

With respect to GNP, the World Bank predicts that it will drop or remain constant in the four countries during the early part of the forecast period (when the countries are continuing to undergo economic reconstruction), but that it will eventually rise by between 5 percent and 6 percent per year in each of the countries. With respect to retail gasoline prices, we expect that they will reflect world market oil prices, which we assume will gradually rise to $30 per barrel by the year 2010. By 1995, we expect that gasoline prices in the Czech Republic and Slovakia, Hungary, and Poland will also reflect tax rates similar to those in Western Europe. We anticipate that retail gasoline prices in Bulgaria will be somewhat lower than in the other three countries.

According to our forecasts of car ownership and car use, both the number of cars and the average annual number of miles traveled per car will increase over the forecast period (see figure 1), but not by a large amount. By the year 2010, there will be more than 13 million cars in the four Central and Eastern European countries—an average of 167 cars per 1,000 people—and each car will be driven an average of slightly more than 5,000 miles per year. The number of cars and the number of miles driven will both be far lower in these countries than they were in the United States and Western European countries in the late 1980s. In fact, because of the decrease in GNP and the increase in gasoline prices in the early part of the forecast period, these numbers will actually fall initially.

Figure 1. Forecasts of the number of cars and the number of miles traveled per car in Bulgaria, the Czech Republic and Slovakia, Hungary, and Poland

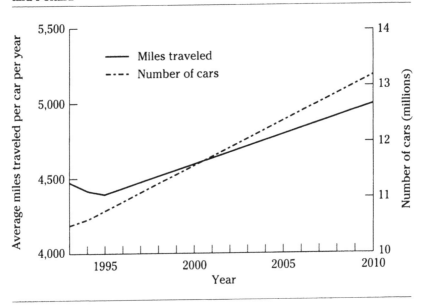

High and low emissions scenarios

To determine how increases in motor vehicle ownership and use will affect levels of motor vehicle emissions of CO, HC, NO_x, and lead in the four Central and Eastern European countries through 2010, we combined our forecasts of these increases with emissions-per-mile estimates obtained by running the MOBILE 5.0 emissions model under two scenarios. In our so-called high emissions scenario, we assumed that the types of vehicles and fuel used throughout the forecast period are similar to those used in 1993. In particular, we assumed that there are some minor controls for HC, CO, and NO_x emissions and that no new vehicles with two-stroke engines are sold, but that no emissions standards are enforced. We also assumed that no effective vehicle inspection and maintenance programs are in place; that all gasoline is leaded; and that vehicles in the four Central and Eastern European countries are older, on average, than vehicles in the West.

In our so-called low emissions scenario, we assumed that new vehicles sold in the Czech Republic and Slovakia, Hungary, and Poland meet European Community (EC)* emissions standards by 1995 and that new vehicles sold in Bulgaria meet these standards by the year 2000. For light-duty vehicles, the EC carbon monoxide emission standard is 4.35 g/mi; the combined hydrocarbon and nitrogen oxide emission standard is 1.55 g/mi. These standards are not as strict as current U.S. standards for vehicle emissions; however, most analysts agree that, in order to meet the EC standards, cars will have to be equipped with catalytic converters. Additional assumptions in the low emissions scenario are that effective vehicle inspection and maintenance programs are put in place; that unleaded gasoline's share of the total gasoline market gradually rises to 80 percent in the Czech Republic and Slovakia, Hungary, and Poland, and to 50 percent in Bulgaria by 2010; that all remaining leaded gasoline has a lead content of 0.15 g/l by the year 2000; and that percentages of younger and older cars in all four countries are similar to those in the United States.

We should note that the most likely emissions scenario might well be one that is intermediate between our high emissions scenario and our low emissions scenario. The high emissions scenario may be pessimistic in that some of the countries in question are already tightening vehicle emissions regulations. For example, all new cars sold in the Czech Republic and Slovakia after October 31, 1993 must be equipped with catalytic converters. Beginning in 1994, all new cars sold in Hungary must meet EC emissions standards. In recent years, the Czech Republic and Slovakia, Hungary, and Poland have all established import policies that favor the importation of new model vehicles and vehicles equipped with catalytic converters. Each of these countries has also set up vehicle inspection and maintenance programs. On the other hand, given the costs entailed by some of the new vehicle emissions control requirements and the

---

*In early 1994, the name of the European Community was changed to the European Union.

poor economic conditions in these countries, enforcement of the requirements might be somewhat lax. For this reason, our low emissions scenario might be overly optimistic. However, both our scenarios should reasonably bound the true level of future emissions, given our forecasts of vehicle miles traveled.

In our forecasts of motor vehicle emissions levels through the year 2010 in each of the four Central and Eastern European countries, emissions are indexed with 1990 (the base year), and emissions in that year are set equal to one. In the high emissions scenario, emissions of each pollutant fall or remain constant until 1995, then gradually rise throughout the remainder of the forecast period (see figure 2) as the annual number of vehicle miles traveled rises. By 2010, total emissions of CO, HC, and NO_x have risen in all four countries. Emissions of CO reach 5,182,000 metric tons, an increase from the base year of 22 percent; emissions of HC reach 930,000 metric tons, an increase of 25 percent; and emissions of NO_x reach 616,000 metric tons, an increase of 5 percent. Lead emissions rise from 2,650 metric tons to 3,040 metric tons, an increase of 15 percent.

In the low emissions scenario, emissions fall continuously throughout the entire forecast period (see figure 3) as a result of new vehicle emissions regulations and the change in the age profile of the vehicle fleet. By 2010, emissions of CO decline to 1,315,000 metric tons, a decrease from the base year of 70 percent; emissions of HC decline to 317,000 metric tons, a decrease of 58 percent; emissions of NO_x decline to 296,000 metric tons, a decrease of 50 percent; and emissions of lead decline to 320 metric tons, a decrease of 88 percent. These figures suggest that that the decreases in emissions per mile that result from the enforcement of EC emissions standards for vehicles, the introduction of strict vehicle inspection and maintenance programs, the decline in the average age of vehicles, and the decrease in the lead content of gasoline far outweigh the expected increase in the number of miles traveled annually.

According to United Nations estimates, some population growth is expected in the four countries—primarily in Poland—over the forecast period; however, it is less than the predicted growth in emissions under the high emissions scenario. Thus, in that scenario there is a slight increase in CO, HC, and NO_x emissions relative to popula-

Figure 2. Forecasts of total motor vehicle emissions in Bulgaria, the Czech Republic and Slovakia, Hungary, and Poland under a high emissions scenario

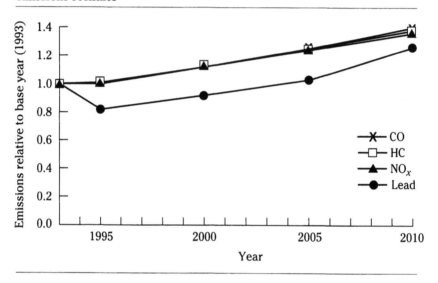

Figure 3. Forecasts of total motor vehicle emissions in Bulgaria, the Czech Republic and Slovakia, Hungary, and Poland under a low emissions scenario

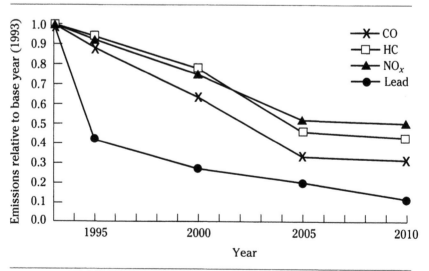

tion. By 2010, approximately 66 metric tons of CO, 12 metric tons of HC, 8 metric tons of NO_x, and 0.039 metric tons of lead are emitted per 1,000 people. With the exception of the figure for lead, these figures are higher than emissions figures for 1990, but they are still far lower than emissions figures relative to population in the United States.

Even under a worst case scenario, our forecasts do not indicate a large increase in total vehicle emissions in Bulgaria, the Czech Republic and Slovakia, Hungary, and Poland before 2010. With the exception of lead, emissions per capita are all less than in the United States. Our forecasts of the number of vehicles on the road and the number of miles traveled per vehicle indicate that it will take several years for the four Central and Eastern European countries to recover from reduced GNP, increased energy prices, and other negative impacts of economic restructuring. By 2010, the level of motor vehicle use in these countries will have risen but will not have equaled that of the United States and most of Western Europe in 1990. Even given relatively high g/mi emissions estimates, this finding suggests that future increases in motor vehicle emissions in the four countries will be small.

Policy implications

The above-noted findings suggest that gradually eliminating leaded gasoline and adopting some other policies to control motor vehicle emissions may be prudent. However, they also suggest that requiring vehicles to be equipped with catalytic converters and setting up vehicle inspection and maintenance programs in order to meet EC standards for emissions of CO, HC, and NO_x may be unwise in the short run. These costly measures may not be needed before 2010 if energy prices are allowed to rise to world market levels and the motor vehicle market is opened up so that relatively new and clean-running vehicles can be purchased in Central and Eastern Europe.

Given that resources are severely limited in Central and Eastern European countries, they might best be devoted to monitoring of ambient air quality and to the identification of cost-effective policies for reducing motor vehicle emissions in cities, where some significant air pollution problems might exist. Such policies could focus on the use of economic instruments that will reduce driving and improve traffic flows, and they could obviate the need to promulgate costly national regulation of motor vehicle emissions.

Margaret A. Walls is a fellow in the Energy and Natural Resources Division at Resources for the Future. She appreciates the assistance of Michael P. Walsh in running the MOBILE emissions model and providing information on vehicles and regulations in Central and Eastern Europe. She also appreciates the helpful comments of Michael A. Toman and the excellent research assistance of Mary Elizabeth Calhoon, Carol Collins, and Eric Nickell.

Assessing the Health Benefits of Improved Air Quality in Central and Eastern Europe

Alan J. Krupnick, Kenneth W. Harrison, Eric J. Nickell, and Michael A. Toman

Assessments of the benefits of improvements in environmental quality in Central and Eastern Europe are needed to determine priorities for pollution abatement activities in the region. A study conducted by researchers at Resources for the Future suggests that the human health benefits attributable to reductions in emissions of three air pollutants in five of the region's countries are potentially large. However, the study also highlights the uncertainties surrounding measurements of decreases in adverse health effects and economic valuations of improved air quality. Attempts to account for these uncertainties yield findings that strengthen the researchers' assertion that air pollution control should be a target of environmental and economic policies in Central and Eastern Europe.

Central and Eastern European countries are simultaneously attempting to address environmental problems and rebuild their economies. Given that substantial financial investment will be needed to accomplish the latter goal, it is important that resources used to attain the former goal be spent on pollution abatement efforts that will garner the greatest benefits to society. Assessments of the benefits that can be obtained through improved environmental quality help policymakers to set rational priorities for environmental cleanups. If such assessments indicate that the benefits are potentially

15

large, they would highlight the importance of not ignoring the environment in pursuing economic restructuring.

We recently conducted a preliminary assessment of one category of benefits that can be obtained from improvements in environmental quality: the health benefits of reduced air pollution. Specifically, we examined the effects on human health of reductions in ambient concentrations of particulates, sulfur dioxide (SO_2), and lead in each of four countries—Bulgaria, Hungary, Poland, and Ukraine. (Some data were also available for the Czech Republic, but the sample clearly was flawed in that it indicated no serious air quality problem anywhere in the country. Data for Slovakia were too limited to be used.)

The assessment has several limitations. One limitation is the small number of pollutants considered. Data availability largely dictated the scope of the assessment. For example, the lack of data on ground-level ozone, which is known to have adverse health effects, meant that the benefits of reducing this air pollutant could not be examined. Another limitation of the assessment is that only one kind of benefit from air quality improvement is analyzed. Scientific uncertainties and lack of data precluded a systematic assessment of benefits other than improved human health that are attained by reducing air pollution. In particular, these uncertainties made it impossible to assess reductions in ecological damages that could result from improvements in air quality. Yet another limitation is that no comparisons are made among the benefits of ameliorating different kinds of environmental hazards. For example, lack of data on the extent and nature of water contamination precluded a comparison of the benefits of improved water quality with those of improved air quality.

Despite these limitations, some important conclusions emerge from our analysis. It appears that reductions in air pollution do have the potential to yield substantial health benefits in Central and Eastern European countries—benefits that are at least 1 percent to 3 percent of gross domestic product (GDP) in these countries and quite possibly equal to an even greater percentage of GDP. While we do not have the information on air pollution abatement costs that would be needed for a full-scale benefit-cost analysis, we believe the size of the potential benefits of reduced air pollution should make air pollution control an important target of environmental and economic policies in Central and Eastern Europe. Control of particulate emissions should be a particularly important target, given that this air pollutant contributes significantly to health damages and is often fairly cheap to abate.

Assessment methodology

Our analysis focuses on the potential benefits of air quality improvements sufficient to meet current European Community (EC) standards for the three pollutants under consideration. Therefore, the first step in the analysis was to establish baseline ambient concentrations of particulates, SO_2, and lead in Bulgaria, Hungary, Poland, and Ukraine. From the World Bank and sources in the region we obtained data on ambient conditions in more than 200 cities and towns, as well as in subdivisions of some large cities (such as Budapest) within these countries. The percentages of total national population that are represented by data on particulates in our sample range from 17 percent (Poland) to 34 percent (Hungary and Ukraine). The percentages of total national population that are represented by SO_2 data in our sample generally range from 19 percent (Poland) to 34 percent (Ukraine) but rise as high as 72 percent (Hungary). Data on ambient lead concentrations were unavailable for Ukraine and were available for only a few urban areas in Hungary and Poland. Therefore, the percentages of total national population that are represented by lead data in our sample range from less than 5 percent (Hungary and Poland) to 23 percent (Bulgaria).

All the data on ambient concentrations of the three pollutants date from the late 1980s and thus do not reflect changes in these concentrations that have resulted from current economic downturns in the four Central and Eastern European countries. Because we do not possess detailed information about the dispersion of pollutants in specific locations, we assumed that all people in a particular sample area have the same pollutant exposures—that is, they all live with the same ambient conditions as those measured at the pollution monitoring stations from which our data are derived.

Table 1. Average percentage reductions in nonattainment areas needed to meet European Community standards for ambient concentrations of selected air pollutants

Country	Total suspended particulates	Sulfur dioxide	Lead
Bulgaria	48	60	23
Hungary	37	12	42
Poland	39	31	23
Ukraine	42	32	not available

After collecting ambient pollutant concentration data, we calculated the degree to which ambient concentrations of each pollutant would have to be reduced in each sample area in order to meet EC standards. EC standards for particulates and for SO_2 limit both average annual exposures and maximum daily exposures. Therefore we calculated reductions in average exposures sufficient to meet both limits (see table1). The percentage reductions we estimated for areas not meeting EC standards are quite substantial, especially for particulates. However, most of the sample areas do turn out to meet the EC standards for SO_2 and lead. (This highlights the problem of nonrandom air quality sampling.)

Our next step was to estimate the reductions in adverse health-related effects that would result if ambient pollutant concentrations declined enough to meet EC standards. We calculated these reductions using a dose-response model that accounts for a variety of health effects, ranging from asthma attacks and so-called restricted activity days to heart disease and premature mortality. The model, which was prepared by researchers at Resources for the Future and elsewhere for a study of the social costs of energy in the United States, reflects a balancing of expert opinions distilled from the clinical and epidemiological literatures on the health effects of air pollution. Two assumptions of the model are noteworthy. The first assumption is that the relationships between doses (exposures) and responses (health effects) are largely linear—that is, the rates of health effects do not grow as exposures increase. The second assumption is that there are benefits from improving air quality even when ambient pollution concentrations are already low.

It should also be noted that the dose-response relationships assumed by the model are based on those documented in the United States and Western Europe. Thus they do not reflect differences in the basic health status of residents of Central and Eastern Europe. We suspect that our model underestimates the reduction in adverse health effects that would occur if pollution declined in the countries included in our analysis. This suspicion is based on the assumptions that individuals living in Central and Eastern Europe are not as healthy as individuals living in the West and that the less healthy an individual is, the more sensitive he or she will be to pollution exposures. Given all the uncertainties and sources of controversy surrounding dose-response relationships, our dose-response model does not attempt to calculate a single response for each health effect. Instead, we assume a probability distribution over the range of potential outcomes for each health effect. This representation of uncertainty is combined with a similar model of uncertainty about the economic value of improved health, as discussed below.

Once we calculated reductions in adverse health effects, we proceeded to calculate a per-unit economic value for the health improvements. Like the preceding step, this step is controversial on both philosophical and practical grounds. Some people are troubled by the notion of assigning monetary values to human health generally and especially to risks of premature mortality. While we recognize these concerns, we believe that it is possible—in principle—to obtain useful information about what trade-offs people are willing to make between health and other social goods.

Even if one accepts the principle of imputing monetary values to health, the practical problem of assessing willingness-to-pay (WTP) for reductions in exposure to health threats must be addressed. Our model for valuing such reductions includes estimates of both direct health damage costs—such as medical expenses and wage rates that reflect the value of workers' restricted activity days—and estimates of WTP for reduced exposure to health threats. The latter estimates are derived from contingent valuation studies in which individuals are asked to reveal their WTP for reduced exposure to health threats. Like our dose-response model, our valuation model attempts to reflect the preponderance of expert opinion in the literature concerning monetary valuations of health effects. It reflects the uncertainties in estimates of WTP through the specification of a probability distribution for each estimate. We then can use so-called Monte Carlo techniques to calculate values of total benefits from air quality improvement that reflect both sources of uncertainty.

All of the WTP and medical cost valuations used in our analysis are derived from analyses of such valuations in the United States and Western Europe. Because we could not develop independent estimates of medical costs and WTP for reduced exposures to health risks for Central and Eastern European countries in the course of our analysis, we adjusted the valuations made in the West to a scale relevant to Central and Eastern Europe. We took two approaches in making this adjustment. The first approach was to scale all values by the ratio of average income in Central and Eastern Europe to average income in the United States—a ratio of about 0.08. This approach may understate valuations of health risks, and particularly valuations of reduced mortality risks, in Central and Eastern Europe. Some evidence suggests that mortality valuations fall much less than proportionally to declines in income. To account for this possibility, our second approach to adjusting valuations was to set the income elasticity for the mortality valuation at 0.35, rather than at an elasticity of 1.0—the elasticity assumed in the relative wage approach.

The final step in our analysis was to estimate aggregate health-benefit values for the countries under consideration. This involved multiplying together the estimated air quality improvement figures—which were derived from ambient particulate, SO_2, and lead concentration reductions sufficient to meet EC standards—and the values of improved health conditions indicated by the dose-response and valuation models. This calculation provides measures of benefits to populations in the sample areas. To obtain benefit figures for the entire population of each country considered, we made assumptions about the pollution to which people not in our sample areas are exposed. We considered two different cases to account for our uncertainty about the pollution exposures of these people. In the first case, the assumption is that the areas not in the sample have air quality that meets EC standards, and thus there is no need to calculate any health benefits for them. This case represents a lower bound for the national benefit figures. In the second case, the assumption is that air quality in the areas outside the sample is equal to the average air quality in the sample areas of each country.

We also considered a case in which all areas in each of the countries under consideration must make the same percentage reductions in ambient concentrations of particulates, SO_2, and lead, whether or not the sample data indicate that the areas meet EC standards for the concentrations. The percentage reductions we set for each country correspond to the average reductions in the sample areas when EC standards are not met. Analysis of this case allowed us to calculate the additional health benefits that could be reaped from making air quality improvements beyond those that would be required to meet EC standards.

Benefit estimates

Our estimates of the health benefits obtained by meeting EC standards for particulates, SO_2, and lead in Bulgaria, Hungary, Poland, and Ukraine indicate that these benefits are potentially large (see table 2). These benefits are expressed as a percentage of a country's GDP in 1988. An examination of the median estimates indicates that the health benefits of meeting the EC standards generally range from 1 percent to 3 percent of GDP, even if we assume that areas not in the sample already meet EC air pollution standards.

If we assume that the air quality of the areas not included in the sample is the same as the average air quality of areas in the sample, the national benefit range would shift to about 5 percent to 12 percent of GDP, given median estimates of health effects and valuations. "Low" estimates of health effects and valuations (the 5th percentile of our probability distribution) shift this benefit range downward, but not as much as "high" (95th percentile) estimates shift it upwards. With low estimates, the national benefit range is about 4 percent to 7 percent of GDP; with high estimates, it is 8 percent to 20 percent of GDP.

The largest part of the estimated benefits are attributable to reductions in ambient concentrations of particulates. Two factors may account for this finding. One is that our data on the percentage

Table 2. Estimates of the economic benefits obtained by meeting European Community standards for ambient concentrations of total suspended particulates, sulfur dioxide, and lead (expressed as a percentage of a country's GDP in 1988)

Country	Estimates for sample population	Extrapolation to national population[1]
Bulgaria	3.0	11.8
Hungary	2.8	11.5
Poland	1.1	6.3
Ukraine	1.7	4.9

Note: Figures reflect middle-range estimates of reductions in health effects and economic valuations of these reductions. Economic valuations reflect valuations made in the West and adjusted to a scale relevant to Central and Eastern Europe on the basis of the difference between the average income in the United States and that in Central and Eastern Europe.

[1] In extrapolating the health benefits enjoyed by a sample population to those enjoyed by a national population, it is assumed that average air quality outside the sample areas within a country is the same as the average air quality inside the sample areas.

of a country's population that is exposed to lead are not as comprehensive as our data on the percentage of a country's population that is exposed to particulates and SO_2. Recall also that most sample areas meet EC lead standards. If we had lead exposure data for Ukraine and more such data for the other three countries under consideration, the benefits of reducing ambient lead concentrations might increase.

Another factor that may account for our finding that the majority of estimated benefits are attributable to reductions in ambient concentrations of particulates is the fact that our dose-response model assigns greater health effects to particulates than to SO_2. Most of the medical literature suggests that particulates are much more harmful to human health than airborne SO_2, which primarily affects materials and ecosystems. However, a portion of SO_2 in the atmosphere converts to sulfate aerosols (SO_4), which are known to be a health hazard but which are measured in the particulate data. Thus our estimates of benefits resulting from the reduction of ambient concentrations of total suspended particulates include benefits ultimately attributable to the reduction of ambient concentrations of SO_2 emissions. It is not possible on the basis of currently available information to determine how much of measured particulates is sulfate aerosols.

Our sensitivity analysis of the valuation of premature mortality risk indicates that assumptions about this valuation significantly affect the outcomes of the valuation. For example, when an income elasticity of 0.35 is used to scale the value of reduced mortality risk in Central and Eastern Europe, median estimates of total health benefits rise to a level comparable to or greater than that when calculations are based on high (95th percentile) estimates of health effects and valuations of reduced health risks. The outcomes of assuming uniform pollution reductions are more mixed. When uniform reductions in ambient concentrations are required across the sample areas, some of which already meet EC air pollution standards, the resulting benefits are greatest in Poland. This country has the largest share of sample locations that meet EC standards, yet it still stands to gain health benefits from additional reductions in air pollution.

Research needs

Our findings support the assertion that air pollution control should be a target of economic and environmental policies in Central and Eastern Europe. Clean air is not a luxury that only rich countries can afford to pursue. Our findings also underscore the importance of controlling particulates—one of the most socially beneficial pollution abatement options.

However, our analysis highlights the large uncertainties we face in putting an economic value on improved air quality. Some of this uncertainty is due to gaps in the basic knowledge of medical science—gaps that might not lessen substantially in the short term. Nevertheless, there are activities that could considerably reduce our uncertainties about valuations of air quality improvement in Central and Eastern Europe. One such activity is to conduct research that will augment knowledge about air quality in the region and the effects of air quality on human health and the environment. Such

research will require intensive data collection efforts and coopera-
tive air chemistry, environmental monitoring, and medical science
research by experts in Central and Eastern Europe and in the West.

An equally important activity is the effort to better understand
the values that residents of Central and Eastern Europe actually
place on improved air quality. Here again, collection of relevant
information about direct damage costs and measurement of willingness-to-pay for improved air quality offers a significant opportunity
for cooperation between experts in Central and Eastern Europe and
those in the West. Although they will be neither easy nor cheap,
such efforts could set the stage for a wide assessment of pollution
damages and priorities. Given the continuing economic and environmental difficulties facing Central and Eastern Europe, these efforts
are an important component in determining responsible environmental policies in the region.

*Alan J. Krupnick is a senior fellow and Kenneth W. Harrison is a
research assistant in the Quality of the Environment Division at
Resources for the Future (RFF). Michael A. Toman is a senior fellow
and Eric J. Nickell is a former research assistant in the Energy and
Natural Resources Division at RFF. The authors gratefully acknowledge
the assistance of the many people who helped them secure and inter-
pret the data used in this study.*

Using Economic Incentives to Reduce Air Pollution Emissions in Central and Eastern Europe: The Case of Poland

Michael A. Toman

A case study of the cost-effectiveness of different policies for controlling air pollution emissions in Poland suggests that the potential benefits of incentive-based (IB) policies, such as emissions fees and emission permit trading, are significant. For example, the study indicates that emissions from large stationary air pollution sources can be controlled at less cost through the use of IB environmental policies than they can be through the use of command-and-control environmental policies. However, the magnitude of the cost savings of IB policies is limited in Poland by several factors. The study also indicates that emission permit trading should be used to complement emissions fees, which are already charged in Poland. However, the legal status of such trading must be clarified.

Like other countries in Central and Eastern Europe, Poland faces the twin challenges of improving its environmental quality while also strengthening its economy during the transition to a market system. One way to reconcile these objectives, as policy analysts have long argued in the West, is to use economic incentives to control pollution. By giving polluters an economic stake in reducing emissions and the flexibility to find least-cost control methods, the argument goes, incentive-based (IB) policies can achieve any specified set of emissions reduction objectives at a minimum total cost to society. To harness economic incentives for pollution control, analysts have advocated the use of emissions fees and the institution of a system of tradable emission permits. A tradable emission permit system establishes a ceiling on total emissions and an initial distribution of allowed emissions among polluters, who can then buy and sell their emission rights. The system thus encourages polluters with the smallest pollution abatement costs to make the greatest pollution reductions.

While the application of such policies in Central and Eastern Europe may seem to be a natural marriage of economic and environmental interests, there are several challenges to consider. Perhaps the biggest challenge is that IB environmental policies are designed to work in countries where polluters actively respond to economic incentives. Even within the context of an established market economy, economic incentives can be distorted by the regulation of product prices and the investment decisions of polluting firms, such as electric utilities. In the emerging market economies of Central and Eastern Europe, the applicability of IB environmental policies is an even more complex issue than in established market economies because the power of economic incentives remains unclear. In particular, continued government intervention in the activities of large state-owned enterprises, which are often the major polluters in a given region, casts a shadow over the use of IB policies.

The application of IB environmental policies poses operational challenges as well. These challenges arise in connection with the monitoring of pollution and the enforcement of pollution standards. Where monitoring is inadequate, even command-and-control (CAC) environmental regulations—in which regulators specify the amounts of pollution individual polluters should cut or the types of pollution control technology to be used—are problematic. By comparison, IB policies require somewhat greater monitoring. To ensure the integrity of emission permit trading, for example, the emissions and permit holdings of individual polluters must be carefully tracked. Legal authority for IB policies must also be clearly established.

In the face of these challenges, two questions arise. First, how large are the potential gains resulting from IB environmental policies compared with those resulting from CAC environmental policies when both types of policies are used to achieve the same environmental goals? If the gains are modest, then efforts devoted to overcoming the economic and institutional obstacles to IB policies might better be spent on addressing other pressing problems. Second, if the gains are worth pursuing, how can these economic and institutional obstacles be overcome in practice?

A study I conducted with Robin Bates of the World Bank and Janusz Cofała of the Polish Academy of Sciences and the International Institute for Applied Systems Analysis in Laxenburg, Austria, attempts to answer these questions. It examines different policies for controlling air pollution emissions in Poland, and it suggests that, in that country, the potential benefits of IB policies appear worth pursuing.

Air pollution control problems and policies in Poland

Our study focused on three main types of primary air pollutants associated with energy use: particulates, nitrogen oxides (NO_x), and sulfur dioxide (SO_2). Each of these pollutants is believed to cause significant environmental damages, although the precise nature and extent of these damages remain unclear. Particulates are known to be a serious human health threat. Sulfur dioxide and nitrogen oxides in acid rain cause ecological damages. In addition, SO_2 converts to atmospheric sulfate particulates that are part of the particulate stream, and NO_x combines with volatile hydrocarbons to produce harmful ground-level ozone.

A large percentage of the above pollutants derives from fossil fuel combustion. The magnitude of the problems caused by the pollutants in Poland is related in turn to the energy intensity of the Polish economy and the lack of effective pollution controls. The energy intensity of gross domestic product (GDP) in Poland is much higher than that in the countries of Western Europe. This intensity reflects the legacy of central economic planning, in which physical production of all commodities—including energy—took precedence over environmental concerns; use of Poland's large endowment of coal, particularly low-grade lignite, was extensive; and economic incentives in the command economy for energy efficiency were lacking. Poland's high energy intensity of GDP and lack of effective pollution controls are reflected in the fact that the ratios of particulate, NO_x, and SO_2 emissions to GDP in Poland are many times greater than those in Western Europe.

While it is commonly believed that environmental degradation is ubiquitous in Poland and other Central and Eastern European countries, in actuality environmental conditions vary considerably across and within these countries. Air pollution in the region of Upper Silesia in southern Poland—and particularly in the areas around the cities of Katowice and Krakow—has been truly dreadful, although it has diminished somewhat as a result of the sharp economic contraction Poland has experienced in recent years. With the exception of areas immediately downwind of particularly dirty pollution emitters, air pollution appears to be at least somewhat less serious in other regions of Poland.

It is also commonly believed that pollution control policies were nonexistent in Central and Eastern Europe until recently. However, Poland charged emissions fees before the political transitions that began in 1989. In 1990, Poland's Ministry of the Environment passed the Ordinance on the Protection of the Air Against Pollution. Under the ordinance environmental standards govern both overall air quality (ambient standards) and the discharges of large factories, power plants, and other large polluters (source standards). Because these standards do not specify the types of technologies that must be used to abate pollution, polluters in Poland enjoy a degree of flexibility in mitigating their air pollution. This flexibility increases the cost-effectiveness of Poland's pollution control system, at least in principle. Since 1989, Poland has also raised its emissions fees (and fines set at a multiple of the regular fees). The increase, which more than offsets inflation, suggests that the fees make some contribution to improving Poland's air quality. Poland's system for monitoring compliance with air quality standards, while far from perfect, is improving.

These observations notwithstanding, there are several problems with Poland's air pollution control policies. First, emissions fees remain low in many cases, indicating that they are more effective in raising revenue (which can be used to ameliorate the effects of past pollution damage or to address other problems) than in inducing polluters to reduce emissions. Second, emissions fees and fines are not enforced in numerous instances because of the practical difficulty of imposing additional costs on enterprises already struggling with economic restructuring. Third, legal issues cloud the application of IB air pollution control policies. Specifically, the Ordinance on the Protection of the Air Against Pollution does not clearly establish the legal basis for emission permit trading among polluting firms. As currently interpreted, the ordinance allows trading among pollution sources within enterprises—among boilers within a power plant, for example. However, it leaves unsettled the scope of trading among polluters. While it has been interpreted as allowing trading among polluters in close proximity to each other, such trading does not appear to be occurring.

Calculating the relative costs of IB policies and CAC policies

In our investigation of the cost-saving potential of IB air pollution control policies, we used a dynamic simulation model that calculates energy use and emissions of air pollution in Poland over five-year intervals from 1990 to 2015 under a variety of emissions control policies. The model, developed at the Polish Academy of Sciences, starts with a scenario reflecting the predictions of experts in Poland concerning economic development and changes in the efficiency of end uses of energy over the period. These predictions give rise to projections of final energy demand—that is, the demand for energy by households, businesses, and other end users of energy—over time. Given these energy demand projections and a specified set of emissions constraints, the model then calculates the least-cost energy supply and energy conversion activities that would be needed to satisfy final energy demand.

Implicit in the use of a model of least-cost final energy supply is the assumption that Poland is moving toward a well-functioning energy market in which producers and consumers will pay the full economic costs of energy (plus any applicable excise taxes). Given the pace and direction of economic restructuring in Poland, this assumption seems reasonable. While the baseline energy demand scenario in the model is fixed, the model does allow for variations in energy demand relative to the baseline scenario. Such variations would reflect the effects of emissions taxes or the cost of emission permits in meeting energy demand.

The model includes a highly disaggregated representation of energy supply and conversion technologies as well as options for air pollution control. With respect to the former, for example, it distinguishes among the various energy production technologies used by power plants. With respect to the latter, it distinguishes among emissions control technologies (such as SO_2 scrubbing) as well as among emissions reduction strategies (such as replacing coal use with natural gas use and constructing power plants that have increased energy efficiency and decreased emissions per unit of fuel use).

Because the model is designed to calculate energy use and air pollution emissions at a national level and because it does not specify the location of specific polluters or how the emissions generated by these polluters affect ambient environmental conditions, it cannot show the effect on ambient air quality of different environmental policies. Nevertheless, we believe that our comparisons of such policies give some indication of the potential benefits garnered by IB policies.

In order to measure the cost-saving potential of IB policies, we established a command-and-control regulation baseline. This CAC regulatory scenario includes emissions standards for large stationary pollution sources that are based on the source standards established by the Ordinance on the Protection of Air Against Pollution. It also includes some controls on coal burning by households (namely, the gradual elimination of this activity in urban areas) and on emissions from the transport sector (most prominently, the installation of catalytic converters in motor vehicles) that are not stipulated in the ordinance.

The CAC scenario and the IB scenarios we compare with it include assumptions of two broad changes that substantially reduce air pollution emissions, quite apart from any environmental initiatives. The first change is a substantial restructuring of the Polish economy, with significant growth in less-polluting sectors (services and light industry) relative to the heavy industry that received undue emphasis in the planned economy. The second change is a dramatic improvement in energy efficiency (by some 35 percent in 2010), consistent with the response to higher market-level energy prices. These changes significantly reduce emissions, even after taking into account the nearly 100 percent increase in Polish economic activity that we assume over the simulation period (see table 1). Thus energy and economic restructuring are the lowest-cost pollution control strategies, since they produce reductions in emissions and an improvement in economic efficiency. The policy comparisons below are concerned with the cost of achieving the further reductions in emissions needed to meet Polish emissions standards.

Table 1. Pollution emissions in Poland case study

Pollutant (thousand tons)	1988	1990	No controls 2010	New controls 2010
Sulfur dioxide	3,827	2,832	3,575	2,017
Nitrogen oxides	1,363	1,187	1,321	906
Total suspended particulates	2,145	1,468	1,211	629

We compared four IB policies with the CAC scenario. The first three of these policies have the same total emissions targets as the policies in the CAC scenario; the fourth does not. The first IB policy retains the CAC scenario's controls on coal burning by households and on emissions from the transport sector, but requires large stationary pollution sources to pay fees on emissions of particulates, NO_x, and SO_2 in order to achieve the total emissions reductions that would be attained under the CAC scenario. The second IB policy relaxes all the CAC scenario's fixed controls on pollutant emissions, including those on coal burning by households and on emissions from the transport sector, and relies on emissions fees paid by large stationary pollution sources and on energy taxes paid by mobile air pollution sources (such as cars) and small air pollution sources (such as homes) to achieve the total emissions reductions that would be attained under the CAC scenario. In order to make the energy taxes equivalent to the emissions fees, the taxes are based on the average volume of pollutant discharges per unit of fuel use. The third IB policy allows large stationary pollution sources to engage in SO_2 emission permit trading at the national level but maintains the CAC scenario's restrictions on particulate and NO_x discharges from these sources. It also maintains the CAC scenario's controls on coal burning by households and on emissions from the transport sector. The fourth IB policy relies on a 100 percent tax on coal use to achieve emissions reductions. In selecting this policy for our analysis, we were interested in exploring the effects of a narrowly targeted tax rather than in attempting to equate the total emissions reductions that would be achieved by the policy with those that would be attained by the CAC scenario.

We compared these four IB policies with the CAC scenario on the basis of social costs (see table 2). These costs include the costs of abating emissions from large stationary pollution sources and the costs of controls on coal burning by households and on emissions from the transport sector. They also include the reduction in consumer surplus that would result from an increase in energy prices and a decrease in final energy demand caused by the pollution control policies.

Results of cost comparisons

Simulations of our model reveal that the application of IB policies to large stationary pollution sources will garner cost savings, but that the magnitude of the savings is limited by two factors that reflect—at least in part—efficiencies already embodied in the CAC scenario. First, as noted above, the CAC scenario's emissions control standards for such sources are based on source standards that do not prescribe the use of particular pollution control technologies. Thus polluters can realize some cost savings under the CAC scenario by choosing the technologies that are least expensive for them. Second, an implicit assumption of our model is that individual firms are free to choose how to distribute pollution abatement efforts among the pollution sources they control—for example, among boilers within a

Table 2. Social costs of pollution control in Poland under command-and-control and incentive-based environmental policies, 1991–2015

Components of social costs	Command-and-control scenario	Incentive-based policies			
		Emissions fees imposed on large stationary pollution sources	Emissions fees imposed on large stationary pollution sources and energy taxes imposed on small and mobile pollution sources	Trading of sulfur dioxide emissions permits by large stationary pollution sources	Tax on coal use
Pollution control costs of large stationary pollution sources and loss of producer surplus	6.57	5.30	5.51	5.87	9.76
Loss of consumer surplus from reduced energy use	0.02	0.02	0.22	0.02	1.04
Costs to urban households switching from coal to gas	0.08	0.08	—	0.08	—
Cost of pollution controls on the transport sector	5.89	5.89	—	5.89	—
Total	**12.56**	**11.29**	**5.73**	**11.86**	**10.80**

Note: All figures represent costs in billions of 1990 U.S. dollars. Each figure represents the total cost of the given component of social costs, discounted at the rate of 12 percent, over the twenty-five-year period between 1991 and 2015.

power plant. This assumption reflects a broad interpretation of the possibilities for intrafirm emission permit trading under the Ordinance on the Protection of Air Against Pollution. Given a narrow interpretation of the ordinance's source standards, firms' flexibility to choose pollution control strategies is limited, and the cost-saving potential of IB policies is increased.

Another implicit assumption of our model leads to an understatement of the cost-saving advantages of IB policies. This assumption is that all the technologies that could be used to abate pollution have already been developed. It does not reflect the fact that IB policies provide dynamic incentives for the development of technologies that would abate pollution at less cost than existing pollution control technologies.

While simulations with our model reveal that the cost savings garnered by the application of IB policies to large stationary pollution sources are limited, they indicate that substantial cost savings can be attained through the use of emissions fees and fuel taxes on all pollution sources and the relaxation of rigid controls on coal burning by households and on emissions from the transport sector. Compared with increased controls on emissions from large stationary pollution sources, the controls on emissions from the transport sector are an expensive means for achieving emissions reductions. Relying only on fuel taxes makes it possible to find a relatively low-cost combination of pollution abatement efforts—most likely, less such efforts by the transport sector and more such efforts by large stationary pollution sources. However, controls on small air pollution sources and mobile air pollution sources in urban areas might still be necessary to satisfy local ambient air quality standards, even if the controls are expensive.

Two other outcomes of the model simulations are noteworthy. First, the emissions fees that are required under the first and second IB policies in order to achieve the emissions reductions attained under the CAC scenario are more than an order of magnitude larger than the emissions fees currently used in Poland. Second, the coal tax (the fourth IB policy) is almost as costly as, but far less effective in reducing emissions than, the other three IB policies and the CAC pollution control strategy. This finding highlights the importance of a comparatively more broad-based emissions reduction strategy.

Implementing incentive-based environmental policies

Our case study of air pollution control strategies in Poland indicates that IB policies can generate cost savings that are, at minimum, nontrivial and possibly substantial; but how can these savings be achieved in practice? Which IB policies are likely to be most effective under the economic circumstances encountered in the transition to a market-based economy?

Emissions fees might be favored in Poland because they already are well established in Polish law and because they generate revenues that can be used to ameliorate pollution or to attain other social goals. However, there are several well-known disadvantages to their use. First, such fees lead to the transfer of substantial revenues from polluters to the government, and polluters therefore oppose raising them. In simulations of our model, emissions fees increased the private cost of compliance with environmental standards by about 75 percent. Second, emissions fees may not have the desired effects in a setting in which enterprises receive budget subsidies from the government and in which the state will either indirectly subsidize emissions fees or not enforce them.

Given these disadvantages, tradable emission permits may be an important complement to the emissions fees currently charged in Poland. A program of nationwide emission permit trading, like the SO_2 emission control program enacted in the United States under the Clean Air Act Amendments of 1990, requires relatively careful monitoring of individual firms' emissions and permit holdings, as noted above, as well as a substantial number of well-informed market participants who are capable of making sophisticated trade-offs. However, less ambitious programs based on a system of bilateral emissions trading also have substantial promise in an economic setting such as Poland's. Under such a system, individual polluters can seek out trading partners in ad hoc fashion (without a formal market mechanism or formal tradable emission rights issued by regulators) in order to find mutually advantageous arrangements. Such arrangements are those wherein a polluter with low pollution abatement costs makes emissions reductions greater than those required by law in exchange for financial compensation from a polluter with high pollution abatement costs. Such arrangements can be conditioned on a requirement that a total emissions reduction goal is attained and that overall air quality in any one area is improved.

Bilateral emissions trading will not result in the exploitation of all possible cost-effective reallocations of responsibility for pollution control. Nevertheless, analyses of comparable trading opportunities in the United States generally indicate that such trading can lead to significant cost savings in pollution abatement efforts. However, these analyses also suggest that the cost-saving potential of bilateral emissions trading tends to be diminished when government restrictions encumber exchanges of permits. Thus the cost-effectiveness of such trading would be enhanced if the legal status of emissions trading were clarified and if government restrictions on exchanges were limited.

There are several potential obstacles to emissions trading in Poland. One obstacle, as noted above, is that the legal status of such trading in Poland is unclear. Another is that incentives to engage in emissions trading would be limited if the government fails to enforce air quality standards or interferes with the economy in ways that weaken firms' interest in minimizing pollution control costs. Under such circumstances, however, any pollution control policy—including emissions fees and CAC regulation—is doomed to failure.

Policy recommendations

Despite the potential obstacles to its success, emissions trading appears to be an important complement to emissions fees in controlling air pollution in Poland. Although such fees stimulate some pollution abatement activities and provide a source of revenue for mitigating environmental damages generated in the past, it is doubtful that they can be raised to the level necessary for Poland to attain its current air quality standards. This point is underscored by the fact that no country in the West has yet managed to raise its emissions fees high enough to rely on the fees to achieve its environmental goals. Thus it seems vital to develop the legal and economic institutions needed to support increased emissions trading. Such trading could start with informal bilateral exchanges, as discussed above, and progress to more formal and multilateral exchanges as Poland's economic and regulatory institutions develop.

It should be noted that the cost-saving potential of emissions trading in Poland might be greater or smaller than our case study indicates. In the study we focused only on how alternative air pollution control policies will affect total air pollutant emissions in Poland. However, the effectiveness of such policies needs to be gauged by how the policies affect actual air quality—that is, ambient pollution concentrations in different locations—as pollution damages depend on ambient conditions. Thus, to improve our understanding of the cost-effectiveness of IB environmental policies in Poland and in other Central and Eastern European countries whose economies are in transition, it is necessary to extend our analysis to an examination of emissions trading in light of local ambient standards and the way pollutants are dispersed as a result of meteorological phenomena. Such an examination will allow us to quantify more accurately the gains from emissions trading under trading rules that reflect how emissions from different pollutant sources affect ambient conditions at different locations. An analysis of this kind by the World Bank has already begun in the Polish city of Krakow.

Michael A. Toman is a senior fellow in the Energy and Natural Resources Division at Resources for the Future. He is especially grateful to Robin Bates and Janusz Cofala, coauthors of the larger study from which this chapter is drawn.

Cost-Effective Control of Water Pollution in Central and Eastern Europe

Charles M. Paulsen

Lack of controls on point sources of pollutant discharges—primarily sewage treatment plants—has contributed to the degradation of surface water quality in Central and Eastern Europe. Neither relying on existing pollution control nor adopting the West's best-available pollution control technology and minimum pollutant discharge policies is likely to be a feasible course of action for the region, as the environmental consequences of the former would appear to be unacceptable and the costs of the latter to be prohibitive. However, a recent case study involving the Nitra River basin in the Slovak Republic suggests that the region can realize substantial improvements in water quality at a fraction of the cost of command-and-control policies used in the West by taking into account the relative contributions to pollution and pollution control costs of individual point sources and basing pollution control efforts on those contributions and costs.

Since political transformations there in 1989, Central and Eastern Europe has increasingly come to realize the severity of the degradation of its surface water quality. Most major rivers and lakes in the region have pollutant concentrations far above international standards. In addition to posing health threats, contamination of the region's surface water has economic consequences. For example, pollutant discharges into the Baltic and the Black seas have already seriously reduced the output of once-productive fisheries.

Policies designed to improve the region's water quality will have to grapple with the declining industrial and agricultural output, concomitant decreases in material living standards, and shortages of investment capital faced by all the region's national governments. Given these conditions, the countries of Central and Eastern Europe could simply choose to delay adoption of the best-available pollution control technology and minimum pollutant discharge policies of Western Europe and North America until their economies can afford them. In the meantime, this decision would mean relying on existing pollution control facilities to deal with water quality problems caused by so-called point sources of water pollution—primarily industrial and municipal sewage treatment plants. As the region's economies improve, presumably more money would become available for the capital investments that are required for construction of sewage treatment plants with state-of-the-art pollution control. The region's governments would meanwhile stand to gain an advantage from delaying investment in water quality improvement: the longer they wait to undertake such investment, the greater the likelihood that noncompetitive industries will fail, obviating the need to invest in new or improved plants to treat the industries' sewage.

Delaying efforts to improve water quality is problematic, however. Although pollutant discharges into the region's waters can be expected to decrease as industries close, change their product mix, or update their production processes, it is likely that municipal sewage loads will increase as more and more households and newly formed businesses are connected to public water and sewer networks. In addition, the public may demand that water quality issues be addressed in the present rather than in the future. The downfall of many of the formerly Communist governments was brought about in part by environmental movements, and anecdotal evidence suggests that a substantial demand for improved environmental quality still exists in many Central and Eastern European countries.

One alternative to delaying water quality improvement efforts would be an immediate attempt to implement a minimum discharge policy, whereby sewage treatment plants would be required to reduce pollutant discharges into surface water in line with European Community (EC) standards for wastewater treatment. However, the cost of such a policy might well be more than governments in the region are willing (or able) to pay, given that the per capita cost of meeting such standards exceeds per capita gross domestic product (GDP) in three of five countries in Central and Eastern Europe (see table 1). Although countries in the region might be able to borrow a portion of the capital investment required to construct new or improve existing sewage treatment plants in order to meet EC waste-water treatment standards, it might not be wise for them to do so. Debt as a percentage of GDP is already high in many Central and Eastern European countries. Moreover, it is likely to increase as investment in industrial modernization and in communications and transportation infrastructure proceeds.

Together, three factors—poor surface water quality, a demand for improvements in such quality, and scarce financial resources—suggest that neither long delays in wastewater treatment nor immediate implementation of a minimum discharge policy is appropriate. If the desire to improve surface water quality and the necessity of minimizing pollution control costs are important factors in decisions made by the governments of Central and Eastern Europe, a policy that attempts to improve water quality cost-effectively would seem to offer a means of realizing the most improvement per dollar invested.

Behavior of pollutants in river basins

Since most of Central and Eastern Europe's water supply is drawn from rivers, these bodies of water can be expected to be the primary focus of efforts to improve water quality. In order to understand which efforts are likely to be cost-effective, it is necessary to take into account two behavioral patterns of pollutants in a typical river basin. To illustrate these patterns, suppose that our typical river basin has three point sources of pollutant discharges and three monitoring stations where water quality is measured, and that point source 1 is located highest upstream, followed further downstream by monitoring station A, point source 2, monitoring station B, point source 3, and monitoring station C (see figure 1). The first behavior pattern to consider is that pollutants from each source of discharges into the basin

Table 1. Resources of and potential costs to improve water quality in Central and Eastern Europe (U.S. dollars)

County	Population (millions), 1992[1]	GDP (millions of dollars), 1992[1]	Per capita GDP, 1992[1]	Per capita cost to meet European Community water quality standards, 1992	Total debt as percentage of GDP, 1991[1]	Percentage change in industrial production, 1990–1992[1]
Bulgaria	8.47	6,903	815	3,755	not available	–54
Former Czech and Slovak Federal Republic	15.66	36,093	2,305	4,927	27	–40
Hungary	10.30	35,494	3,446	2,116	78	–32
Poland	38.30	72,579	1,895	1,230	61	–32
Romania	23.20	14,152	610	1,422	not available	–54

[1] Figures are from The Economist (March 13, 1993)

[2] Figures are from Der Standard (June 17, 1993).

Figure 1. Typical river basin

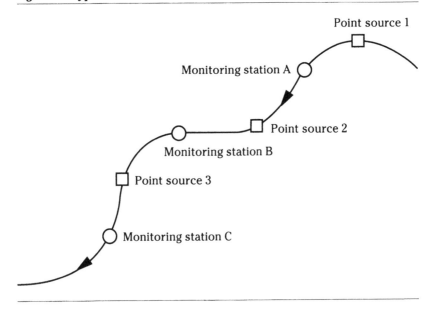

move only in a downstream direction, resulting in higher quality of water upstream and lower quality of water downstream. Thus the quality of water passing by monitoring station C will be affected by pollutants discharged from point sources 1, 2, and 3, while the quality of water passing by monitoring station A will be affected only by pollutants discharged from point source 1. The second behavioral pattern to consider is that most conventional pollutants—nitrogen and phosphorus, for example—either decay naturally and so are effectively removed from the river as they move downstream or settle out of the water column and become deposited in the sediment of the river bed.

The downstream movement and the natural decay or settling out of conventional pollutants in rivers have several implications for management strategies to enhance water quality. First, even if all point sources of a pollutant discharge the same quantity of the pollutant into our typical river basin and cost the same amount to control (an extremely unlikely circumstance), the relative importance of each point source with respect to improving water quality at the critical monitoring station will differ. If the worst water quality is found at monitoring station A, only the control of discharges from point source 1 would make any contribution to improving water quality. If, on the other hand, the worst quality water is found at monitoring station C, control of discharges from point sources 1, 2, and 3 would contribute to water quality improvements. In the latter case, it is likely that discharges from point source 3 will have far greater effects on water quality at monitoring station C than will discharges from point source 1. Thus the location of point sources makes a difference in the effects of the point sources on water quality at various places in the river basin.

It is particularly important to consider differential effects on water quality due to the location of point sources when the financial resources needed to reduce pollutant discharges are scarce. When this is the case, an analysis of potential pollution control policies that accounts for the location of pollution sources along our typical river basin may be needed to identify the policy that will meet ambient water quality targets in the most cost-effective way. If water sampled at monitoring stations B and C meets such targets, while water sampled at monitoring station A does not, a policy that attempts to achieve the requisite pollution control at least cost would focus on controlling pollutant discharges from point source 1. In the more likely case that water sampled at monitoring station C has the worst ambient quality, environmental authorities would need information on the relative contributions of all three point sources to water quality degradation, as well as on the relative costs of controlling discharges from each of the sources, in order to construct a policy that meets ambient quality standards at least cost. The basic idea is that the more a point source contributes to environmental degradation, the more it should control its pollutant discharges. Similarly, the less it costs a source to control its discharges, the more the source should control discharges relative to other sources.

Nitra River basin case study

A study I conducted with László Somlyódy of the International Institute for Applied Systems Analysis in Laxenburg, Austria, suggests that Central and Eastern Europe might be able to improve its ambient water quality substantially by considering the relative effects and pollution control costs of point sources of pollutant discharges into river basins and to do so in a way that would be cheaper than adopting the minimum discharge and best-available technology policies of Western Europe and North America. The study of alternative water quality enhancement policies accounts for the location of each major point source of discharges into the Nitra River basin, which is located in a heavily industrialized area of the Slovak Republic; the pollution control costs of each of these sources; and the effects of each source's discharges on the basin's ambient water quality. The study focuses on concentrations of dissolved oxygen, which are often used as a broad measure of the quality of water and the health of aquatic ecosystems, and it considers the effects of three types of policies to increase such concentrations. The first policy is to require point sources to increase the current concentration of dissolved oxygen in the basin by reducing pollutant discharges to the lowest possible level using the best-available pollution control technologies. The second policy is for the region in which the Nitra River basin is located to implement what for it would be the least-cost strategy for increasing the concentration of dissolved oxygen in the basin to 4.0 milligrams per liter (mg/l), a concentration high enough to sustain fish and other forms of aquatic life. The third policy is for the region to implement what would be the least-cost strategy for increasing this concentration to 6.0 mg/l.

A comparison of the costs of each of these three policies reveals that the minimum discharge/best-available technologies (MD/BAT) policy is the most expensive (see table 2). While this policy would increase the concentration of dissolved oxygen in the Nitra River basin to 6.9 mg/l, it would do so at an annual cost of approximately $14.4 million (U.S. dollars). In contrast, the annual cost of each of the least-cost policies is less than half this figure. The least-cost policy to increase the concentration of dissolved oxygen to 6.0 mg/l would entail an annual cost of $6.6 million; the least-cost policy to increase this concentration to 4.0 mg/l would entail an annual cost of only $2.8 million. Both least-cost policies represent a substantial improvement over maintenance of the sta-

Table 2. Comparison of base case and alternative policies to increase the concentration of dissolved oxygen in the Nitra River basin

Policy	Minimum concentration of dissolved oxygen (mg/l)	Annual cost (millions of U.S. dollars)	Percentage of cost of MD/BAT policy
Maintain status quo (base case)	0.7	0	not applicable
Minimum discharge/best-available technologies (MD/BAT)	6.9	14.4	100
Regional least-cost (4 mg/l)	4.0	2.8	19
Regional least-cost (6 mg/l)	6.0	6.6	46

tus quo (the base case), even though the cost of the latter is zero. This is because the currently low concentration of dissolved oxygen in the basin—0.7 mg/l—is likely to be detrimental to many forms of aquatic life.

The above cost comparisons illustrate the likely ratio of cost savings that could be achieved through the use of a least-cost policy to increase concentrations of dissolved oxygen. The 4.0 mg/l concentration could be achieved at less than 20 percent of the cost of the MD/BAT policy, while the 6.0 mg/l concentration could be achieved at less than 50 percent of the cost of this policy. The question that arises is whether similar cost savings would be realized if least-cost pollution control policies were applied to river basins larger than the Nitra River basin. Given the magnitude of potential cleanup costs relative to GDP in Central and Eastern Europe, the answer to this question is doubtless of considerable interest to the region's governments.

Adoption of least-cost pollution control policies

Despite the fact that resource economists have been advocating their use for more than two decades, least-cost pollution control policies are the exception rather than the rule in practice. Although the United States has recently adopted one such policy—trading among electric power plants of permits to emit sulfur dioxide—it and many other countries in the West have traditionally made little attempt to design and implement pollution control policies that are efficient in the sense that they will lead to ambient standards being met at the lowest possible cost. There are many reasons why such policies are not promulgated more often. They include technical difficulties in projecting the economic and environmental effects of alternative policies, concerns about whether pollution control costs will be evenly distributed among pollution sources, and the lack of institutions to coordinate management of environmental resources.

Given that cost-effective pollution control policies are not the norm in the West, it might be expected that Central and Eastern European countries would be hesitant to adopt them. However, these countries' severe resource constraints and their institutional flexibility—the result of recent political transformations in the former Soviet bloc—tend to make such policies particularly attractive and potentially easier to implement than in the West. This combination of conditions suggests that Central and Eastern European governments may be more attuned to the arguments of resource economics than Western governments have been to date.

Charles M. Paulsen is a fellow in the Quality of the Environment Division at Resources for the Future.

Tradable Sulfur Dioxide Emission Permits and European Economic Integration

Dallas Burtraw

An international system of tradable emission permits has engendered interest as a way to control emissions of sulfur dioxide (SO$_2$) in Europe. Like other forms of incentive-based regulation, emission permit trading has the potential to achieve a given emissions reduction goal at less cost than command-and-control regulation. However, the full cost-saving potential of SO$_2$ emission trading in Europe's electricity industry, which generates 65 percent of Europe's total SO$_2$ emissions, is significantly undermined by the structural and regulatory diversity of that industry. Despite this fact, emission permit trading can be justified on the ground that it would promote the reflection in electricity prices of the social costs of pollution resulting from electricity generation. The internalization of social costs in these prices is critical to Europe's realization of economic unification and to the liberalization of European energy markets.

Emissions of sulfur dioxide (SO$_2$) are thought to contribute to acidification of soils and water, deterioration of visibility, and corrosion of materials; they are also thought to aggravate respiratory problems in humans. One possible approach to the control of SO$_2$ emissions is the trading of SO$_2$ emission permits on an international level. In 1990 the United States initiated the largest experiment of this type of environmental regulation in history by adopting a national system of SO$_2$ emission permit trading.

A system of tradable emission permits typically works as follows. A cap is set on total emissions of a given pollutant in a given geographic area, and a fixed number of permits to emit some quantity of the pollutant is issued to the pollutant emitters, who can then trade the permits among themselves. In theory, those who find that their pollution abatement costs are relatively low will choose to reduce their emissions and sell their unneeded permits, while those who find that their pollution abatement costs are relatively high will choose to buy permits rather than to reduce their emissions. Ideally, decisions about whether or how much to invest in pollution abatement and in permits will depend on the relative cost of each.

Two potential benefits are often ascribed to a system of tradable emission permits. The first is that such a system allows an environmental objective to be achieved at least cost. The second is that, because the price of emission permits tends to be reflected in the price of final products and services, individuals are encouraged to consider the social costs of pollution in their consumption decisions.

The first of these benefits will be difficult to obtain through the trading of SO_2 emission permits within Europe's electricity industry, which is a major focus of SO_2 emissions control efforts in Europe. The lack of incentives to minimize costs and other obstacles to emission permit trading in that industry suggest that the cost savings of such trading may not be fully realized. Nevertheless, the fact that SO_2 emission permits can lead to the internalization of social costs in the price of electricity provides a compelling reason to pursue emission permit trading.

SO_2 emissions reduction and electricity markets

Among the most important efforts to curb SO_2 emissions in Europe to date are a protocol issued by the United Nations Economic Commission for Europe (UNECE) in 1985 and a directive adopted by the European Community (EC) in 1989. The protocol—which was signed by twenty-one countries, not including Poland and the United Kingdom, and ratified by sixteen countries in 1987—calls for UNECE member countries to reduce their annual level of SO_2 emissions by at least 30 percent of levels in 1980. The Large Combustion Plant Directive specifies percentage reductions in emissions of sulfur dioxide and nitrogen oxides for each EC member country and places specific limits on such emissions from power plants.

According to researchers at the International Institute of Applied Systems Analysis (IIASA) in Laxenburg, Austria, the combined effect of current SO_2 emissions control commitments on the part of European countries will be to reduce Europe's total annual SO_2 emissions in 1995 by perhaps 29 percent of levels in 1980. However, a reduction of 50 percent to 70 percent of 1980 levels must be attained in order to reverse soil acidification caused by sulfur deposition.

By the year 2000, sulfur dioxide emissions are expected to decline significantly in the northern and western regions of Europe and to decline somewhat less significantly in the countries of Central and Eastern Europe. However, these emissions are expected to increase in the southern region of Europe, the countries of which have not yet committed to any SO_2 emissions reductions targets.

The attention given SO_2 emission reductions by many European countries reflects a concern about environmental protection in general. In turn, this concern has been reflected in economic unification efforts. For instance, the Single European Act (SEA) of 1987 placed protection of the environment on an equal footing with economic growth, free trade, and policies to encourage competition. The Treaty on Political Union has further strengthened political consideration of the environment in the EC. As both the treaty and the SEA suggest, the EC—which has replaced member states as the initial source of environmental regulations—recognizes that economic growth as a consequence of economic unification will have environmental implications.

With regard to reducing SO_2 emissions, the question at hand is how the EC's commitment to environmental protection will affect Europe's path toward economic integration and vice versa. Since approximately 65 percent of all SO_2 emissions in Europe originate from the generation of electricity, the structure and regulation of European electricity markets will be an especially important factor in determining how SO_2 emissions reductions will be achieved. Historically, financial subsidies have played an important role in

European markets for electricity and other forms of energy. These subsidies have cost European taxpayers billions of European currency units (ECUs) over the years. The EC has claimed that the buying and selling of electricity among EC member countries could produce cost savings of between 1.5 billion and 3 billion ECUs per year and that an internal market for all forms of energy could lead to savings of 0.5 percent to 1 percent of gross domestic product across the EC. An open international electricity market is expected to be developed in several phases. In the market, EC member states will retain authority for electricity system planning and all aspects of electricity pricing.

Benefits of tradable emission permits

For two reasons, tradable emission permits have attracted more interest than emission fees and other forms of incentive-based (IB) environmental regulation as a means for regulating SO_2 emissions in Europe. First, under an emission permit trading system, the EC could directly control the total level of SO_2 emissions in Europe by restricting the number of permits it issues to SO_2 emitters. Second, by constraining trading in certain areas or by certain polluters, it could control the level of such emissions in the locations that suffer the most from the environmental and health effects of sulfur deposition.

As noted above, there are two potential benefits of an emission permit trading system. The most commonly cited benefit is the attainment of productive efficiency—that is, the cost-effective achievement of a given goal. Emission permit trading (and other forms of IB environmental regulation) can often achieve a given environmental goal at less cost than command-and-control (CAC) regulation of emissions. Unlike CAC regulation, under which regulators might specify use of the best-available pollution control technology, emission permit trading encourages the development of both production technologies and pollution control technologies that in the future would reduce the cost of complying with emissions limits set by regulators.

The second potential benefit of emission permit trading is that it helps to internalize in the prices of goods and services the costs to society of pollution emitted after firms have complied with environmental laws. Under CAC regulation, this residual pollution remains unpriced because firms are given free-of-charge access to the land, air, or water that assimilates it. As a consequence, society demands too much electricity because the price of electricity is too low. Under an emission permit trading system, firms must pay through the purchase of emission permits for access to the environmental media that assimilate their residual pollution. The resulting pricing of residual pollution raises product prices and causes secondary adjustments in economic behavior, such as decreases in consumer demand, that promote efficiency in the allocation of resources to economic activities.

Obstacles to emission permit trading

Simulations of SO_2 emission permit trading in Europe conducted by IIASA and others indicate that such trading could greatly reduce the costs of meeting Europe's SO_2 reduction goals. One opportunity for reducing these costs stems from the fact that EC member states often pursue pollution abatement strategies that subsidize important domestic constituencies but that do not obtain pollution reductions at least cost. Another opportunity for reducing costs stems from the considerable differences across the EC in preexisting emission standards that form the benchmark for percentage emissions reductions required by the Large Combustion Plant Directive. The disparities in marginal abatement costs produced by these differences could be exploited in a system of emission permit trading. Under such a system, a country could determine whether it would be cheaper to reduce its emissions by a given amount or to compensate another country for reducing its emissions by the same amount, and act accordingly. Unfortunately, neither of these two opportunities to reduce costs is likely to be exploited in the implementation of emission permit trading in Europe's electric industry.

Virtually all analyses of the potential savings from emission permit trading and other forms of IB regulation have been limited to examinations of competitive product markets or to the role of market power on behalf of producers in an unregulated product market. These analyses depend fundamentally on the expectation that firms will respond to economic incentives by choosing a least-cost strategy for compliance with environmental standards. With regard to the electricity industry in Europe, this expectation may not be realistic for several reasons.

European electric utilities may not have the incentive to maximize profits or to minimize costs that competitive firms have. State-owned electric utilities might lack incentives to minimize costs for several reasons. First, doing so might cause their production targets to be increased by the government in the future. Second, they are protected by the government from financial failure through subsidies, tax exemptions, easily obtained credit, and other forms of financial aid. Thus their survival and growth may depend more on their relation to the current government bureaucracy and on certain aspects of their performance that are of concern to society than to success in the market.

Privately owned electric utilities might also lack incentives to minimize costs. With the exception of those in England, almost all privately owned utilities in Europe are lightly regulated monopolies that typically recover all their costs through tariffs. Because they are not subject to prudence reviews that can lead to the disallowance of some costs, they can engage in pricing that approximates cost-plus pricing, in which firms pass along all costs to consumers. In some instances, such as when prices are based on standard rather than on actual costs, they will have a modest incentive to reduce costs.

There are two other reasons why the potential cost savings of emission permit trading are unlikely to be realized. First, firms that are regulated typically do not make decisions on the basis of market prices, but rather on the basis of distorted opportunity costs that reflect regulatory practices. Implicit in the current regulatory practices of European countries are biases in the treatment of depreciation, recovery of capital costs, risk associated with investments, and fluctuating production input prices. Asymmetries among European countries in the nature of regulation and in incentives to pursue specific environmental compliance options will tend to undermine the economic principles of an emission permit trading system. Second, regulators in each country are likely to favor explicitly certain types of compliance activities that promote social objectives other than the control of pollution at least cost.

Investment behavior on the part of all electric utilities is affected by the national energy policies of European governments, which have played an increasingly important role in planning electricity-generating systems and in making decisions about new investments in the electricity and other energy sectors of their national economies since the early 1970s. Energy policies have been and continue to be used as a tool for macroeconomic policies and for policies aimed at providing aid to specific fuel industries. For example, energy policies in Spain, England, and Germany support the use of coal; those in the Netherlands support the use of gas; and those in France support the use of coal and nuclear energy. Such policies clearly discourage the implementation of efficient cost control practices.

One way to overcome, at least in part, the lack of incentives to minimize costs in the electric industry is to promote the EC goal of making electricity pricing transparent—that is, distinguishing among the various components of delivered electricity services and explicitly accounting for the costs embedded in each component. Transparent pricing of electricity has two benefits. First, it discourages subsidization of electricity generation technologies, because the costs of the technologies are open to public review and criticism. Second, it promotes competition in electricity markets by helping to establish the relative cost advantages of technologies.

In the United States an element of price transparency is provided by the Uniform System of Accounts, which recommends accounting practices to state regulators. As the EC moves toward an internal energy market and increased economic integration with the rest of Europe, the establishment of a similar institution in Europe would promote emission permit trading by helping to make assymetries in the regulation of electric utilities apparent. A template for consistent cost accounting in the electric industry should be developed as part of an international agreement for SO_2 emission permit trading across Europe.

Emission permit trading in an international economy

Despite the many obstacles to minimizing emission reduction costs through emission permit trading, such trading should remain of interest because it has a significant virtue. It and other forms of incentive-based environmental regulation allow decision makers to consider social costs in the context of their own opportunity costs and, depending on how product prices are set, to reflect social costs in product prices. In internalizing social costs in the price of electricity, emission permit trading is consistent with the economic objectives of the EC—free trade and economic competition—and with the economic integration of Europe. However, strategic considerations in the implementation of international environmental policies may undermine this virtue of emission permit trading. These considerations suggest that international environmental agreements must not only specify environmental goals but also articulate the mechanism through which such goals are to be achieved.

Consider an international agreement for SO_2 emissions control that specifies (potentially tradable) emission targets for each country but leaves the mechanism for achieving the targets up to the individual country. Is there reason to believe that, acting unilaterally, national governments would implement national systems of tradable emission permits? A country's firms may find it cheaper to comply with such IB regulation than to comply with CAC regulation, but their cost savings must be weighed against the reduction in their international competitiveness that would result from the effect of emission permit trading on their product prices. This effect, as noted above, is to raise product prices by including in the prices not only the marginal cost of pollution control but also the opportunity cost of using the atmosphere's absorptive capacity. The distinction between marginal and average costs makes a direct comparison of emission permit trading and CAC regulation in Europe difficult, but it is probable that a firm's opportunity cost in purchasing an emission permit—a cost reflected in the monetary value of the permit—would be higher than the firm's savings in environmental compliance costs under emission permit trading. In this case, the national benefits that result from competitive pricing under CAC regulation might outweigh the benefits that result from savings in pollution control costs under IB regulation. Furthermore, the benefits to individual firms that utilize comparatively more polluting technology might be greater than those to firms that utilize comparatively less polluting technology.

Hence, absent a specific mechanism to implement IB environmental regulation as a component of international environmental agreements among European countries to reduce SO_2 emissions, it is unlikely that national governments of Europe will unilaterally adopt such regulation. As a result, the costs of achieving environmental goals may be higher than they need to be, and the necessary precepts for the economic integration of Europe—namely, transparent pricing of electricity and the elimination of subsidies implicitly provided by some national environmental policies—may be undermined.

Given these potential consequences, it makes sense for the countries of Europe to pursue a system of tradable SO_2 emission permits on an international level. A critical issue in implementing such a system will be the initial distribution of emission permits. In order to realize cost savings through the exploitation of marginal emissions abatement costs, the permits should be distributed directly to electric utilities and other SO_2 emitters rather than to national governments. If permits are initially allocated to nations, there is no guarantee that the price of a permit or the opportunity cost of discharging a unit of SO_2 emissions would be passed on to industry or to consumers. Furthermore, it is likely that national governments would use CAC regulation that is consistent with their holdings of permits to achieve emissions standards set under an international environmental agreement. If so, governments would have an opportunity to subsidize domestic electricity utilities and other domestic industries that make extensive use of electricity.

One of the benefits of emission permit trading is that the allotment of permits can promote cost-sharing among countries that each bear different emissions control costs. Within the electric industry, the fuels used to generate electricity are an important factor in determining these costs. While there is great diversity among European countries in the fuels used to generate electricity, there is little diversity in the fuels any one country uses to generate electricity. The countries whose electric utilities rely on coal in generating electricity will bear the greatest costs to reduce SO_2 emissions. To lessen the cost burden of these countries, permits could be distributed on the basis of historic levels of emissions. However, this approach has one disadvantage. Given current regulatory practices, electricity prices would be unlikely to reflect the opportunity cost of using permits if the permits were distributed to electric utilities free of charge. An alternative approach is to distribute permits through an auction. However, coal-based electric utilities would oppose this approach. Thus it may be preferable to distribute the majority of permits as endowments, doing so on the basis of historic levels of emissions, and to auction the remaining permits. Over time the endowments could be phased out and replaced by an expanding auction. Revenues from permit auctions could then be allocated in ways that would lessen the burden of emissions control on the electric industry as a whole and on the countries bearing the greatest emissions control costs.

Importance of internalizing social costs

An international system of tradable emission permits will not be easy to implement in Europe. Economic integration of Europe is unlikely to lead individual countries to surrender control of their energy, environmental, and industrial policies. Moreover, each country is likely to continue to appease influential social interests within its borders.

Additional problems will lessen the prospects for achieving the benefits of SO_2 emission permit trading in Europe's electricity industry. Structural and regulatory diversity within this industry is an obstacle to realizing the full cost-saving potential of emission permit trading because it imposes regulatory biases that obscure the full social opportunity cost of tradable SO_2 emission permits. As noted above, the adoption of accounting practices that make this cost explicit and the distribution of emission permits to individual electric utilities rather than to national governments would help solve this problem.

Even if the full cost-saving potential of tradable emission permits cannot be realized in Europe's electricity industry, the internalization of social costs in electricity system planning and potentially in the price of electricity is sufficiently critical to advancing Europe's agenda of economic integration and liberalization of energy markets that it alone justifies emission permit trading. Such trading and other forms of IB regulation are the only kind of environmental regulation consistent with the economic objectives that have been set out in Europe. However, the lack of incentives for the countries of Europe to implement unilaterally SO_2 emission permit trading at a national level is an obstacle to the internalization of social costs in energy prices. In order for such internalization to occur, international negotiations on transboundary pollution must establish emission permit trading (or some other form of IB regulation) as the mechanism for achieving SO_2 emissions reduction goals.

Dallas Burtraw is a fellow in the Quality of the Environment Division at Resources for the Future.

The Allocation of Environmental Liabilities in Central and Eastern Europe

James Boyd

Existing soil and groundwater contamination are likely to affect future industrial development and investment in Central and Eastern Europe, because large-scale pollution cleanup costs are potentially tied to industrial property transactions in that area of the world and the division of liability for these costs is uncertain. Determining how pollution cleanup costs should be allocated between governments and current or future property owners will not be easy. Retroactive liability is unlikely to be a desirable or a feasible means of assigning such costs for several reasons—one reason being its costly impact on the large number of thinly capitalized firms in the region. Publicly financed liability funds, by widely distributing cleanup costs, create a more desirable climate for foreign and domestic investment than does a U.S.-style system of retroactive liability and will lead to better setting of priorities for cleanup efforts. However, pooled fund programs should be operated on a short-term basis only, as they may reduce private incentives to invest in pollution reduction.

Privatization and market reform in the economies of Central and Eastern Europe are occurring against a backdrop of severe environmental degradation left by decades of inadequate government attention to environmental conditions. The legacy of soil and groundwater pollution inherited by the new governments of Central and Eastern Europe not only creates direct health and ecological costs but also is likely to affect future industrial development and investment. The contentious development and expensive implementation of legal and regulatory approaches to mitigating environmental degradation in the United States and the European Community (EC) suggest that environmental problems in Central and Eastern Europe—where economies are much weaker, environmental problems much greater, and legal and regulatory institutions much less developed than in the West—will only be resolved at great economic and political cost.

In theory, an effective environmental liability system in Central and Eastern Europe would serve to deter the future generation of pollution by threatening polluters with liability costs arising from improper waste generation or disposal. However, a more immediate, practical consequence of new liability rules is the assignment of responsibility for existing pollution. This assignment raises an important question—namely, how should the costs of removing or reducing existing pollution be allocated between governments and current or future property owners? The answer is complicated by the financial weakness of both governments and property owners in Central and Eastern Europe, the costs and uncertainty involved in the quantification of environmental risks, the political nature of liability reform, and the need to promote domestic and foreign investment.

While the costs of remediating existing pollution in the former Soviet bloc cannot be precisely estimated, they are clearly huge. The estimated cost of meeting EC or U.S. environmental standards in Poland, for example, is as high as $300 billion. The costs of remediating existing pollution are highly uncertain, due to the lack of lending and insurance institutions familiar with risk assessment and to the virtual nonexistence of accounting, zoning, or regulatory requirements for documenting risk-generating processes or technologies. Large-scale pollution costs are thus potentially tied to transactions involving industrial property in Central and Eastern Europe.

Uncertainty regarding potential liabilities is exacerbated by the lack of established liability concepts, legal precedent, and consistent enforcement principles in Central and Eastern Europe. The formerly communist countries have no common law traditions—such as those in the United States—that allow environmental claims based on concepts such as nuisance, trespass, negligence, or strict liability. Instead, they use a civil law approach that imposes damages almost exclusively in cases where there has been a violation of a government standard or regulation. A civil law, rather than a common law, definition and enforcement of liabilities presents an opportunity for governments to coordinate and achieve cost-effective resolutions to the cleanup of existing pollution. However, it also creates uncertainty for potential investors. Because such a system is defined neither by precedent nor by a consistent application of judicial principles, the scope and division of potential liabilities are unclear.

The unique environmental and institutional conditions in the countries of Central and Eastern Europe argue for liability approaches that may differ from those advocated in countries with more advanced legal systems, less pollution, and greater economic vitality. Given these unique conditions, the environmental liability systems established in Central and Eastern Europe should be influenced by two goals. First, in light of the need for economic growth, liability rules consistent with the promotion of privatization and foreign investment should be favored. Second, because Central and Eastern European governments lack the funds to pay the entire cost of cleaning up existing pollution, legal and regulatory policies should be designed to target public revenues toward the environmental hazards that pose the greatest threat.

To pursue these goals, liability initiatives in Central and Eastern Europe should distinguish between the timely and effective implementation of liability rules governing the creation of future environmental risks and the efficient cleanup of pollution generated in the past. These are entirely different issues. The first concerns the question of how to create incentives for future pollution reductions, while the second concerns the question of how to efficiently achieve a distributional goal—that is, how the costs arising from pollution created in the past should be borne. With respect to the latter, it can be easily argued that both moral and legal responsibility for existing pollution lies primarily with the former Soviet bloc governments themselves. The question of who should bear the costs of remediating existing site contamination is particularly important since it is likely to affect patterns of foreign and domestic investment and clearly affects the value of initial asset endowments distributed in the process of privatization.

The argument against strict and retroactive liability

One liability approach that might be instituted in Central and Eastern Europe is strict and retroactive environmental liability. This type of liability holds the current owner of a property fully liable for pollution cleanup and compensation costs, even when the pollution was generated by past owners or users of the property. While strict and retroactive liability strongly deters the future generation of pollution, its application in the United States has prompted debate over its inequitable allocation of responsibility for cleanup costs and its potentially adverse impact on property development.

Independent of judgments about its effects on pollution reduction and economic activity in the United States, strict and retroactive liability is unlikely to be desirable or even feasible in Central and Eastern Europe for several reasons. First, given the weak condition of the economies of Central and Eastern Europe—which is due in large part to capital scarcities—such liability could impose costs high enough to force many domestic producers to declare bankruptcy or liquidate their assets. Given the magnitude of existing environmental hazards, the full internalization of costs based on a strict and retroactive application of liability might yield negative real asset values for a significant fraction of industrial properties. These consequences are inefficient, since bankruptcy and asset dissolution involve costs in the form of abandoned capital, lost firm-specific human capital, and reduced competition. In any event, shortages of capital and the tenuous financial position of newly privatized firms suggest that liability rules dependent on firms' ability to liquidate or otherwise free capital to compensate for environmental damages will be ineffective.

Second, a strict and retroactive liability system is not likely to lead to effective priorities for cleanups. Under such a system, only the most unpolluted properties would be sought for development, and resources would therefore be devoted to the cleanup of relatively unpolluted properties.

Third, a strict and retroactive liability system will likely stifle foreign investment, which is critical to the acquisition of skills, technology, and capital by Central and Eastern Europe. Foreign investors' concerns about retroactive liability derive from their experiences with huge retroactive liability costs in domestic markets, the fact that their firms' capital is relatively available to be tapped in the event of liability actions, and the lack of political stability—and hence investor influence—in Central and Eastern Europe.

Fourth, the political and ethical "polluter pays" motivation for strict and retroactive liability does not in general apply in Central and Eastern Europe. Because former governments and managers of cooperatives are most to blame for existing pollution, there is little ethical justification for the new owners of privatized properties to be liable for the past sins of others.

Fifth, the distributional impact of strict and retroactive liability poses a threat not only to the success and timeliness of cleanups of existing pollution but to the success of liability rules aimed at future pollution reduction. As in the West, environmental policies in Central and Eastern Europe will be derived and enforced in a political context, and their distributional impacts will largely determine their legislative and political success. The fact that liability rules have large distributional, and hence political, consequences can influence the evolution and enforcement of environmental pollution law. Because the profitability (or existence) of new enterprises is potentially threatened by strict and retroactive liability, resources will be directed at the political system to redistribute costs. One natural way to do this is to seek changes in the liability rules themselves. Given the political context in which liability laws are formulated, it follows that rules dealing with future liabilities should be separated from rules dealing with retroactive liabilities in order to enhance political acceptance of a tough prospective liability system. A more equitable distribution of the costs arising from existing pollution makes laws aimed at future pollution reduction more politically and economically sustainable.

From a practical standpoint, however, separating the costs of existing pollution from the costs of pollution being generated by new property owners is difficult, since precise definitions and divisions of responsibility for pollution require costly risk assessment efforts. Even in the United States and EC, an initial, noncomprehensive environmental audit can cost hundreds of thousands of dollars for a major industrial site. Because the risk assessment capabilities of Central and Eastern European governments and industries are greatly inferior to those of their counterparts in the West, it is unlikely that an accurate division of responsibility for pollution is possible at a reasonable social cost. When existing pollution is widespread but difficult to detect with conventional site assessment methods at the point when ownership of a property is transferred, and when advanced risk assessment technologies and expertise are in short supply, a precise technical—let alone legal—separation of responsibility for pollution cleanup may be unrealistic.

Contractual mechanisms for allocating retroactive liability

Given uncertainties regarding the scale of and the liabilities implied by existing pollution, different contractual mechanisms to allocate liability may be needed to improve the efficiency of privatization and foreign investment decisions. The desirability of alternative contract forms is largely a function of the type of information available and the point in time at which information is acquired. With respect to the latter, assessment of liabilities may occur either before the fact (ex ante), at the point of transaction, or after the fact (ex post), following the transaction.

Knowledge of either responsibility for or the extent of pollution may be available to only one of the parties to a property transaction. For example, a government may have knowledge of existing pollution risks but choose not to reveal them prior to a transaction. On the other hand, if the government is unable to ascertain when pollution was generated and grants amnesty for retroactive liability, an investor might inflate the value of risks he or she claims to have inherited at the point of sale.

In the unlikely event that both the buyer and the seller have complete knowledge of all the risks posed by pollution on a property, there are two primary contractual possibilities. One is for the seller (the government) to guarantee that no liability will be assessed for existing hazards. The other is for the government to impose strict and retroactive liability on the buyer but discount the price of a property to account for the costs of such liability. The virtue of the latter approach is that the property transaction would be immune to "renegotiation" by the government. Therefore, subsequent disputes over which hazards did or did not exist at the point of sale would be avoided.

When the buyer cannot observe contamination of a property ex ante, he or she is purchasing an asset of unknown quality. Given this, an optimal contract requires insurance against levels of risk that differ from those revealed by the seller. Should the seller know that the property is clean, that person can simply guarantee to compensate the buyer for expenses resulting from any subsequently revealed contamination; alternatively, he or she can absolve the buyer of liability. If contamination created before the sale can be separated from that created after the sale, an optimal contract would release the buyer from retroactive liability costs. Having done so, the asset price would reflect the property's value net of retroactive liabilities. When the buyer cannot observe contamination ex ante, retroactive liability for the buyer is clearly not desirable, since the costs of liability are not known at the point of sale and so cannot be accurately discounted from the asset price.

If pollution generated by the buyer cannot be separated from pollution existing at the point of sale, however, a liability amnesty would give the buyer a loophole to escape the costs of the pollution he or she generated. In this case, government assumption of liability may be inefficient. The conflict between buyer uncertainty over liability and the creation of loopholes by liability amnesties underscores the importance of environmental audits, which allow for an accurate separation of responsibility for pollution.

If a clear separation of responsibility is not practical, the question that remains is how to distribute the costs of existing pollution while creating incentives for the reduction of future pollution generation. An imperfect but potentially desirable approach is to provide relief from retroactive liability through the provision of government funds earmarked for cleanups. Two distinct forms of liability funds exist. One form is pooled funds, which provide public money for cleanups and compensation—money that would be provided by property owners in a Superfund-type liability system. The other form is the liability escrow account, in which a fraction of a property's purchase price is set aside and earmarked for cleanup costs defined at a later date. A crucial difference between the two funds is that escrow funds provide money to clean up pollution at one specific property, while pooled funds provide money for the cleanup of any number of properties.

Pooled funds

Pooled funds, which are conceptually related to "no-fault" pollution insurance, have been instituted or proposed in several countries to deal with large-scale environmental risks. They can differ in terms of duration, limits on the nature and scale of costs covered, and criteria—such as compliance with regulatory standards—that must be met in order for property owners to be eligible for reimbursement of cleanup costs. In all cases, however, only a fraction of liabilities are borne by property owners, with the balance being borne by the pooled fund.

Pooled funds are contrary to the notion that the polluter should pay cleanup costs. However, the use of public moneys for cleanups in Central and Eastern European countries is more easily justified than in western countries, since decades of state ownership and central planning in the former imply that responsibility for existing pollution lies largely with the governments of Central and Eastern Europe.

Because they widely distribute the costs of environmental cleanups, pooled funds may represent the least economically disruptive mechanism for dealing with large retroactive liabilities—liabilities that could otherwise force the abandonment of properties or the bankruptcy of property owners. The administration of such funds allows for the coordination and rationalization of a nation's risk reduction activities. With centralized control of the system, pollution mitigation measures at sites presenting the greatest social risks could, in principle, receive priority. The caveat is that an effective, centralized system of risk identification and ranking does not currently exist and is difficult and costly to implement.

Firms that expect large retroactive liabilities to ultimately force the closure of their enterprises pose a particularly serious pollution threat. If the enforcement of liability is delayed due to an overburdened legal system or the slow pace of regulators in identifying pollution sources, such firms have no incentive to reduce pollution in the period before enforcement occurs. Faced with the likely prospect of closure due to existing property contamination, such firms will find it profitable to pollute at will until the government forces them to cease their operations. A benefit of publicly provided cleanup funds, then, is that they increase the expected value of the firms, reduce the likelihood that firms will close as a result of retroactive liability costs, and thus lead firms to make investments in pollution control based on the now-realistic ability to continue profitable operation well into the future. Since legal and regulatory enforcement of liability claims is likely to take some time in Central and Eastern Europe, this benefit of publicly provided cleanup funds is particularly important.

There are potentially significant problems associated with the use of pooled funds, however. Because the current owners of properties where cleanups are to be conducted will be reimbursed for the costs of cleanup, one problem is that price competition in the market for risk remediation services may be lessened unless the government polices a bidding process for such services. The market for environmental cleanup services, while sure to become increasingly competitive, is not competitive. Another problem is that property owners have an incentive to engage in costly cleanup activities that might have little social benefit because the owners are, in effect, insured by the government. Limited fund levels combined with inflated remediation costs could swiftly deplete a pooled fund's reserves.

A more fundamental problem associated with pooled funds derives from the inability to adequately separate risks existing before a fund is set up from those created during the fund's lifetime. If pooled fund programs cannot effectively distinguish between retrospective and prospective sources of pollution, amnesty for retroactive liability would carry a significant danger—namely, that newly created hazards will be claimed as hazards created in the past, thus allowing property owners to escape liability for pollution they generated. Realistically, a complete assessment of retroactive liabilities in Central and Eastern Europe will take years. Thus, while pooled funds may have desirable short-term benefits, it is unequivocally undesirable for them to become a permanent fixture in a government's environmental policy portfolio. If firms believe that all or even a fraction of their potential future liabilities will be borne by a subsidized government fund, the private incentive to invest in pollution reduction would be reduced. Thus an important political question is how a pooled liability fund program can be effectively operated on a short-term basis and then phased out to create liability assignments that effectively internalize environmental costs.

Escrow funds

In some Central and Eastern European countries, the government allows part or all of the proceeds of a property sale to be set aside in escrow. These escrow funds can then be used for the cleanup of pollution on the property. In practice, escrow funds guard against the incentive problems created by a pooled, no-fault liability fund by placing clear limits on the amount of public funding that will be provided for cleanups. In addition, because they typically expire after a period of months, escrow funds limit the ability of new property owners to escape the costs of pollution that they generate in the future.

Like pooled funds, escrow funds provide some insurance against existing pollution costs for property purchasers. Compared with the alternative of strict and retroactive liability, such insurance discourages a government from setting a costly standard for cleanups once it sells a property. The reason is that costly cleanups deplete the escrow funds—which would revert to government coffers if not drawn down.

However, new property owners may have an incentive to deplete escrow funds as much as possible, since for them the funds represent a source of costless pollution remediation financing. The result is that government funds that could be used to address pressing environmental problems might instead be used to address relatively unimportant environmental problems. Investors will clearly seek to purchase the least polluted properties. If funds are dedicated to the further cleanup of these relatively clean properties, government revenue that could be used to reduce environmental risks at relatively polluted sites would be reduced. As compared with escrow funds, pooled funds provide the government greater authority to determine the allocation of cleanup funds.

The creation of an effective liability system

Both pooled fund and escrow fund systems will create a beneficial form of insurance against pollution liabilities, and thus will stimulate foreign and domestic investment and potentially smooth the transition to a tough future liability system. However, public funding of pollution remediation should be viewed as only a short-term means of addressing existing soil and ground-water cleanup issues in Central and Eastern Europe. Any permanent government subsidy of soil and ground-water cleanups will only continue to distort private property owners' incentives to reduce pollution.

Pooled funds present the best opportunity for targeting public funds to the cleanup of pollution posing the greatest health and ecological threats. However, they also represent a form of subsidy that might be politically difficult to dismantle. The challenge for the governments of Central and Eastern Europe, then, is to provide public funding for pollution remediation, but in a way that leads private property owners to believe that in the future they will be responsible for the social costs of polluting activities. While it may be tempting to give generous liability amnesties to foreign firms in order to encourage investment, doing so may lead foreign firms to export environmental risks to Central and Eastern Europe. Limitations on existing liabilities are desirable, but there is nothing to recommend investment incentives created by weakened liability rules that are aimed at reducing future pollution generation.

The strict enforcement of private property rights (and the assignment of liabilities) is not a particularly effective way—in the short term—to guarantee a rational social approach to pollution reduction. The reason is that private interests pursued through a liability system need not coincide with the social interest when resources are severely limited. The economies of Central and Eastern Europe are not currently robust enough to support large resource expenditures aimed at the resolution of legal disputes.

It remains the case that the costs of existing pollution must be distributed in some fashion. Moreover, a system of incentives for future risk reduction is desperately needed if current environmental conditions are to improve. Pooled liability funds are likely to be the most politically and economically effective mechanism for distributing costs and reducing risks. However, they should be subject to safeguards—specifically, a limit on the duration of coverage and requirements for eligibility to claim reimbursement of cleanup costs (such as the installation of pollution reduction technologies). Pooled funds are a necessary compromise between strict and retroactive liability and unrealistic attempts to perfectly and quickly separate responsibility for existing and future pollution.

James Boyd is a fellow in the Energy and Natural Resources Division at Resources for the Future.

Environmental Policies, Economic Restructuring, and Institutional Development in the Former Soviet Union

Michael A. Toman and R. David Simpson

Foreign aid in the form of technical assistance might be useful in help-ing the newly independent states of the former Soviet Union deal with past and current pollution. But such aid probably will not have a lasting, positive impact in the absence of reforms in the countries' basic social institutions. Without development of the institutions of a market econo-my, environmental measures are unlikely to be successful. Obstacles to investments that promote economic and environmental improvements must be removed if these improvements are to be achieved.

Technical and regulatory efforts to improve the management of environmental quality in the former Soviet Union are a focus of programs to provide foreign assistance to the newly independent states that once made up that country. To assess the prospects for the success of these assistance programs, policymakers in the United States and other Western countries must address many basic questions about the new states' environmental policies and their transitions from planned to market economies. These ques-tions fall into three categories.

First, what can we expect regarding the investment incentives of firms in these countries? How many low-cost investments that im-

73

prove both the environment and the economy will the firms undertake, and why are these investments not already being undertaken?

Second, how will the restructuring of enterprises and institutions alter the responses of polluters to environmental policy instruments? Conversely, how do the challenges of restructuring that face enterprises affect the design of environmental policy?

Third, how will environmental policies and fiscal policies interact? How will environmental policies interact with industrial and overall social security policies? For example, where enterprises that are not viable in the long run are being temporarily maintained on social grounds, what investments should be pursued to reduce enterprise losses and environmental damages?

Based on our observations in Russia and Ukraine, and on extensive discussions with experts in those countries, we believe that real progress on environmental problems in the countries of the former Soviet Union will lag until there are substantial and far-ranging reforms in basic economic, legal, and social institutions. We do not deny that some targeted technical assistance could produce substantial improvements in environmental quality and quality of life for individuals affected by the assistance. Without basic institutional reforms, however, it is doubtful that these countries will have the capacity to continue the progress made possible by foreign assistance and to generate substantial environmental improvement on their own.

Our justification for this conclusion goes beyond the observation that the states of the former Soviet Union remain poor and that their resources available for environmental investment remain limited. It also goes beyond the observation that, in the absence of development assistance, excessively strict environmental regulation likely will be politically unpalatable.

Given these countries' current social institutions, it will be costly, if not impossible, to succeed in translating a public demand for environmental improvement into concrete action. The necessary political, legal, and economic accountability needed to do this effectively does not yet exist. Moreover, even if there were agreement on the need for change, institutional failures in the economy would likely raise the cost of enforcing environmental standards well above even the levels experienced under inefficient command-and-control programs in otherwise functional market economies.

These observations in turn raise doubts about the cost-effectiveness of major environmental assistance programs in the former Soviet Union without significant institutional reform there. Evidence is growing that improvement of environmental quality is a highly valued objective in the countries in question. However, environmental quality, as well as economic performance, might be better served first by assistance that helps the countries of the former Soviet Union to develop the institutions of a market economy, including the associated legal institutions of property, liability, and contract law.

Environmental policies in Russia and Ukraine

Environmental policies in Russia and Ukraine illustrate both the disarray in the environmental policies of the countries of the former Soviet Union and the difficulties in improving these policies without progress toward the development of market economies. Environmental policies in these countries consist of a hybrid of standards for emissions and fees on emissions in excess of the standards. Environmental regulators tax pollution at two rates: all emissions are subject to a low tax rate, but emissions in excess of standards set for each source are subject to a higher rate. In principle, environmental regulators also have the power to order polluters to reduce emissions or to cease operations if the emissions pose a serious threat to public health or ecological integrity.

Emissions standards are based on essentially arbitrary distinctions among hundreds of pollutants. Far more standards exist than regulators can monitor or enforce. Standards also are set rigidly for individual sources of emissions, without regard for differences among the emissions reduction costs for each source or for differences among the impacts of each polluter on actual pollution concentrations. Regulators express concern that flexibility in the ways polluters are allowed to comply with emission standards, as with emission permit trading, would expand the opportunities of firms to exceed their emissions allowances—although current rules already require that emissions sources be monitored. This concern is ironic (even surrealistic), given the current scale of pollution violations.

In principle, these problems could be lessened by overhauling environmental regulations. However, other shortcomings in environmental regulation reflect economic and social concerns, as well as environmental concerns, and thus are harder to address.

The administration of pollution fees in Russia and Ukraine is problematic in several respects. Because expenditures for environmental protection are financed mostly by fees on polluters rather than from general revenues, environmental regulators are faced with a fundamental contradiction: to address environmental concerns arising from one set of activities, they must tax pollution from other, quite possibly unrelated, activities. If regulators were to charge pollution fees high enough to encourage substantial reductions of pollution, they would risk undercutting the tax base.

Revenue raising largely motivates the setting of emissions standards. To maintain tax revenues, the government often sets emissions standards a few percentage points below prevailing emissions levels. If the system worked to reduce emissions, it would require a ratcheting upward of standards to raise revenues, compromising firms' incentives to make long-term investments in environmental improvement. However, pollution fees simply are too low to achieve much environmental improvement, especially for state enterprises that do not face the normal budget constraints of a market economy. Moreover, at current rates of inflation, increases in pollution fees are rendered negligible shortly after they are announced.

With fees having little impact on pollution, the only other line of defense is legal sanctions against egregious violators of pollution standards. In practice, however, the problems of economic and political transition render this option largely ineffectual as well. Because so much of the economy in the former Soviet Union remains under state control, attempts to enforce environmental sanctions become intramural conflicts among government ministries. In this situation, the rule of law with regard to the environment often is quite weak, especially in light of the strong vested interests in maintaining enterprise operations that we discuss below.

Aside from problems related to the administration of emissions standards and pollution fees, the allocation of funds for pollution cleanup and reduction projects leaves something to be desired. The determination of priorities for environmental expenditures is not necessarily linked to environmental benefits. Some effort to identify such benefits is made when different expenditure proposals are considered. However, an important criterion for allocating funds appears to be the financial need of the local government or enterprise proposing a pollution cleanup or reduction project. Thus funds are often made available for projects that local governments or firms cannot finance on their own, with little regard for the benefits the projects generate by reducing serious health or environmental risks arising from pollution.

Obstacles to making win-win investments

In the long run, the industrial pollution problems of the former Soviet Union can only be overcome by major investments in more efficient and cleaner production processes and equipment. Many such investments probably could be undertaken at very low cost and result in both substantial environmental benefits and lower production costs. The existence of such "win-win" investments begs an important question, however: if such options are available, why have they not been pursued? Some of the reasons that these possibilities continue to be unexploited may be traced to the Soviet legacy, others to the difficulties of the transitional period, and still others to problems of information and oversight common, to greater or lesser degrees, in all economies.

One impediment to win-win investments is the morass of regulation and licensing requirements left over from central planning. These requirements make starting new businesses and instituting substantial reforms in existing ones extremely difficult. As a result, they discourage the establishment or retooling of firms that are both more profitable and less polluting.

A second part of the Soviet legacy that impedes win-win investments is the tradition of propping up faltering firms with public funds. This tradition undermines incentives for both increased efficiency and pollution reduction. If enterprise managers know that they will be bailed out with public funds, regardless of the performance of their firms, they have little incentive to seek cost-saving production innovations. Moreover, they may have little incentive to adopt even low-cost solutions to their environmental problems if they believe that these solutions will sooner or later be financed out of public funds.

One ongoing impediment to win-win investments is obstacles to both foreign and domestic private investment. Such investment is limited by several factors. First, the process of privatization has just begun in Russia and is even less advanced in many other states of the former Soviet Union. Second, the institutions that characterize capital markets and the banking system in Western economies are just now coming into being in these countries. For example, corporate law is very incomplete, and accounting procedures that would enable outside investors to determine the value of potential investments have not yet been adopted. Third, taxes on the profits of firms in the countries of the former Soviet Union are substantial. These taxes, along with exchange controls and high inflation (which triggers high interest rates), limit the attractiveness of new investment.

A second ongoing obstacle to win-win investments is the limited capacity of the labor market to adjust to the transition from planned to market economies. Labor mobility remains limited because workers have traditionally obtained all social services (including housing) through the enterprises that employ them and because internal migration is subject to state control. Another circumstance that has made it difficult for the labor market to adapt to the transition is the financial stake employees have in some newly privatized enterprises. Share ownership in such enterprises is largely concentrated among workers and managers, increasing the financial losses that employees would face if these enterprises fail. Because employees realize that they face risks of financial loss and unemployment in emerging market economies and they distrust the prospects for success with new investment, they are exerting political pressure for their governments to subsidize firms or to take other measures to prevent firms from failing.

A seemingly simple solution for overcoming the obstacles to win-win investments would be to allow firms to sink or swim on their own merits. However, this solution may not be feasible under the current circumstances of the transition from planned to market economies. These circumstances may combine to deprive even deserving enterprises of the financing they will need to survive. Thus, in deciding which otherwise failing enterprises receive financial aid and regulatory leniency, decision makers in the countries of the former Soviet Union must distinguish between firms that make obsolete products using archaic production methods and those that are trying to make needed products using newer, cleaner production methods. Moreover, given restrictions on the social "safety net," decision makers need to take into account social factors that may outweigh considerations of narrowly defined economic efficiency in making these decisions.

Thus the limitations on private activity to improve the environment and the economy at the same time also hinder governmental policies for pollution control. While the possibility of publicly funded bailouts exists, polluting firms will be unresponsive to economic sanctions. Moreover, legal or economic sanctions that threaten employment, the viability of enterprises, and the social fabric will be vigorously opposed by enterprise managers, employees, and the branches of government that still oversee polluting industries. As long as public and private decision makers remain unaccountable for their decisions, firms' capacity to change their environmental behavior is very limited.

Institutional reforms

Environmental policies per se probably will have relatively little effect until there is progress toward greater general strengthening of economic and legal institutions. The development and maturation of the institutions of capitalism in the former Soviet Union may be facilitated by increased macroeconomic stability, the establishment of legislation to govern corporate conduct and reduce regulatory barriers to the creation of new businesses, the reform of financial markets, and the revamping of the provision of public goods and social security.

While the transition from planned to market economies proceeds, it is important that environmental policy move in tandem with the general development of economic, legal, and social institutions. For example, while some flexibility in approaches to the enforcement of environmental regulations is called for, it is also important that decision makers be able to predict the effects of these approaches. Actions that would further decrease the confidence of potential investors could be counterproductive, even if they achieved some short-term environmental improvement.

It is also important that the institutions of environmental policy reflect the changing technical capabilities of regulators, the evolution of judicial and other institutions, and the increased stability of firms. Case-by-case reviews of compliance strategies should be replaced by general regulations that incorporate flexibility in compliance. Incentive-based measures, such as limited emissions-permit trading programs, should be established and expanded as opportunities arise. A more concerted effort to set priorities for environmental expenditures and to limit soft enterprise budgets (budgets that are based on the expectation of publicly provided funds to make up losses) would probably improve the efficiency of environmental expenditures while the economy as a whole makes the transition toward greater private financing of environmental improvements. Finally, some simplification of the environmental standards themselves would be beneficial.

Building better economic and political institutions is time consuming and does not offer the immediate and tangible rewards that technical support systems may afford. However, institutional reforms are crucial if the technical support programs for environmental improvement that are now being championed in foreign assistance debates are to be successful.

Michael A. Toman is a senior fellow and R. David Simpson is a fellow in the Energy and Natural Resources Division at Resources for the Future.

Economic Restructuring and the Environment: Exploiting Win-Win Opportunities

Michael T. Rock

Governments in the newly democratic states of Eastern Europe and the former Soviet Union are under pressure to deal with serious environmental problems at the same time that they are making the difficult transition from centrally planned to market-based economies. Relying on newly created environmental ministries to solve these problems is probably a mistake; instead, the governments of Central and Eastern Europe should focus their immediate environmental policies on the "win-win" opportunities created by the economic transition—opportunities to improve economic conditions and the environment at the same time. Pursuing these opportunities will require, as a first step, the maintenance of political and economic stability. It will also require development of the right incentives with respect to foreign trade and a shift to market-oriented agricultural production. The pursuit of win-win opportunities is no guarantee of long-term environmental sustainability, but it will allow breathing space in which to create environmental legislation and a cost-effective public-sector environmental regulatory institution, both of which are prerequisites for such sustainability.

Most of the former socialist countries in Central and Eastern Europe are attempting simultaneous transitions to democracy and to economic policies that expand the role of markets and trade. These double transitions have confronted policymakers with enormous challenges. No country, prior to 1989, ever abandoned social-ism for democratic capitalism. Consequently, there is no direct evi-dence on how to transform a centrally planned economy with Len-inist political structures into democratic capitalism. It is also not clear how much existing theory can act as a guide. Economists and others disagree over the appropriate timing and sequencing of reforms and over the relative importance of pursuing legal/institu-tional reform, more efficient resource allocation, and more rapid technical innovation.

Lack of clarity regarding the basic economics and politics of a simultaneous political and economic transition makes it difficult to consider how to deal with the environmental legacy of communism. However, serious hazards to human health need to be addressed. Moreover, because environmental groups played a pivotal role in the collapse of communism, they have come to exert substantial influence over subsequent political developments in Central and Eastern Europe. This influence burdens newly democratic govern-ments with heavy environmental responsibilities at a time when they are severely limited in their abilities to finance improvements in environmental quality.

Governments in each of the newly independent states of Eastern Europe and of the former Soviet Union have attempted to address their environmental problems by relying on the extremely limited capabilities of newly created environmental ministries. I believe this is a mistake, at least in the short term. Instead, I would propose focusing environmental policy during the transition on the "win-win" opportunities created by the economic transition—opportunities to improve economic conditions and the environment at the same time. These gains can be reaped by a variety of economic reforms described below.

Options for reform

Despite economic travails and the weaknesses of existing environ-mental institutions, governments in Central and Eastern Europe can act to improve environmental quality immediately. But they will need to be highly selective in what they do and how they do it. How should they go about this? They should consider the environmental consequences of their economic transitions. While there is no definitive conclusion yet, the weight of existing evidence on the relationship between economy-wide policies and the environment suggests that what is good for market-oriented growth often is good for the environment as well. This evidence also suggests that one point of departure for economic policy reforms should be changes that enhance the efficiency of resource use, increase economic growth, and improve environmental quality.

Focusing on these win-win opportunities is a good starting point for at least three reasons. First, these opportunities generate economic gains that can be used for improving environmental quality, among other things. Second, they conserve on the use of extremely limited administrative capacities. Third, they limit the need to mobilize new resources for environmental protection in declining economies.

Getting the basics right

Pursuing win-win opportunities first requires focusing on getting basics right—including political stability, stability in the institutions that undergird the economy (specifically legal, corporate, and financial institutions), and macroeconomic stability.

The last forty years' experience with economic development in the Third World has demonstrated that political stability of whatever ilk, as well as stability in the institutions that support a country's economy, are underrated determinants of economic growth. It takes years of both before businesses begin to trust that the stability is likely to be maintained. It takes additional time before businesses decide that investing on the basis of the legal, corporate, and financial institutions that underpin a country's economy might be worth a try. And it takes even more time before businesses decide on products and plant design, complete trial runs, correct mistakes, and eventually succeed. If political stability is lacking or if legal, corporate, and financial institutions are not stable for long enough, this process is short-circuited, and growth and development do not occur.

Political stability and stability in economic rules are also requisites of good environmental policy. Without them, good governance and rule of law will not prevail and neither governments nor private-sector actors can be held accountable for their actions, including their environmental actions. Lack of stability in legal (as well as corporate and financial) institutions also limits the flow of new investment, depriving countries of access to more efficient and cleaner technologies.

It is not enough that the economic rules be stable; they must also promote high savings and investment rates and channel investment into high-productivity areas, not just encourage "rent-seeking" behavior. Doing this requires, among other things, basic macroeconomic stability. Low and stable inflation rates, sustainable fiscal and current account deficits, and competitive exchange rates have been essential to rapid economic growth in the past.

Macroeconomic instability adversely affects the environment in several ways. First, rising inflation rates and growing foreign exchange shortages disrupt production. Environmental quality rises to the extent that production falls in polluting industries, but it does so at the expense of employment, income, and output. While falling production may be environmentally beneficial, it is obviously not a long-term solution. Moreover, the economic crisis attending macroeconomic instability could undermine popular and government support for environmental protection, making it more difficult to protect the environment later.

Second, macroeconomic instability places great strain on government expenditures. Because newly created environmental ministries tend to have limited bureaucratic clout within government, this often leads to serious underfinancing of those ministries, making it difficult (if not impossible) for them to monitor and enforce environmental standards.

Macroeconomic instability—alone or combined with a lack of clarity regarding legal concepts related to the assignment of responsibility for environmental damage caused in the past by former state-owned enterprises—can slow privatization and stifle the flow of new domestic and foreign investment. The stalling of privatization and investment can, in turn, deprive countries of access to more efficient and cleaner technologies.

Trade and the environment

Developing the right incentives with respect to foreign trade is also extremely important to economic growth. Although there is some doubt about the impact of trade policy on this growth and on income distribution, many economists believe that export-oriented or, at least, policy-neutral trade policies promote faster and more equitable economic growth than "import-substitution" industrial (ISI) policies, which seek to develop and maintain domestic production capacity even though imports might be cheaper.

How would a shift to more open trade and industrial policies affect the environment in the former communist countries? Critics contend that it can only lead to more environmental degradation. In practice, the effect of trade liberalization on the environment depends on a cluster of interrelationships—between trade policy and growth, on the one hand, and between growth and environmental degradation, on the other. It also depends on the impact of trade policy on the composition of output, on the use of natural resources, and on industries' decisions about where to locate. While not definitive, a range of evidence now suggests that open trade policies may be more environmentally friendly than the alternatives. For one, growing evidence suggests that it is possible to de-link economic growth from pollution.

Compared to the alternative (ISI policies), open trade policies contribute to this de-linking by reducing the pollution intensity of output at all levels of income per capita. ISI policies, by encouraging an industrial structure that places greater emphasis on heavy (and dirtier) industry at the expense of lighter (and cleaner) industry, have been shown to increase the pollution intensity of output. This deleterious environmental effect is alleviated by trade liberalization.

Countries following ISI policies often subsidize the use of energy and fossil fuels. By shifting from these policies to export-oriented trade policies with their more efficient pricing of energy and fossil fuels, it should be possible to reduce environmental degradation by reducing the energy intensity of output. Recent World Bank studies of the former socialist economies in Central and Eastern Europe suggest this energy-intensity effect could be substantial. In countries with export-oriented trade policies, the pollution intensity of output can be further reduced by a significant transfer of environmentally friendly technology from Western countries.

Trade policy can also affect the environment through its impact on natural resource use. Critics of open trading policies argue that an export orientation encourages unsustainable natural resource mining, especially in countries facing severe foreign exchange shortages. A broad range of evidence now suggests that unsustainable mining of the resource base is not primarily the result of trade policy. Other factors—government policy failure in a specific resource sector (forest policy, for example), landholding patterns, land tenure insecurity, and population pressures on the land—exert more influence.

Finally, trade policy might affect the environment through industrial location patterns that encourage "pollution haven" effects. However, there is little systematic evidence to suggest that dirty industries in rich countries with more stringent environmental standards move to poor countries with less stringent standards. The absence of such effects has been attributed to many factors: the already low tariff rates in rich countries; relatively low pollution control and abatement costs, even for the dirtiest industries; and the importance of wage rates, exchange rates, and political risks.

Agriculture and the environment

The future importance of agriculture in the restructured economies of Central and Eastern Europe is unclear. However, agriculture is likely to be an important component of economic activity in at least some countries, such as Ukraine. Several policies could improve the efficiency of the agricultural sector in the region. In addition to price reforms that make agriculture more productive, these policies include public-sector investments in rural infrastructure, such as roads, irrigation systems, electrification, credit institutions, delivery systems for high-yielding seeds and fertilizer, agricultural extension, and agricultural research.

Critics of these policies contend that they contribute to environmental degradation through soil erosion and runoff of agricultural chemicals, as well as to negative human health effects due to the inappropriate handling of these chemicals. However, there is no simple relationship between policies to improve the efficiency of the agricultural sector and the environment. This relationship depends strongly on the way in which changes in cultivation practices are carried out. That is, the impact of agriculture on the environment is highly dependent on soil conditions, the slope of areas planted, crop choices, and tilling and pesticide-use practices.

This is particularly true in the former socialist economies, where large-scale, mechanized agriculture has been commonly practiced. Extensive soil erosion there resulted from inappropriate tillage practices; the lack of environmentally benign plowing, planting, and tilling equipment; and economically inefficient cross subsidies that encouraged farming on ecologically marginal and fragile lands. Overexploitation of the land also resulted in the loss of nutrients from soils. This overexploitation, in turn, has been associated with the spread of agriculture to fragile dry lands, to a focus on increasing output at all costs, and to the lack of clear property rights in land. Finally, inappropriate handling and application of agricultural chemicals resulted from poor application equipment, poor pricing policies, and nonexistent or ineffective policies for chemical storage and handling.

How might a shift to market-oriented farming affect the environment? Experience elsewhere suggests that clarifying property rights in agriculture can contribute to investments in land that slow both soil erosion and "nutrient mining." Getting prices right for agricultural inputs and outputs helps to ensure that fertilizers and pesticides are applied at appropriate rates. Both of these reforms should stimulate the demand for environmentally benign tilling and chemical-application equipment. Reducing the heavy cross-subsidies of socialist agriculture should shift crop cultivation away from marginal and fragile lands. Taken together, this package of reforms should reduce the negative environmental consequences of existing agricultural practices.

Reducing the environmental damage of these practices will require more than getting prices right, however. It will require getting institutions right, particularly those for agricultural research and extension. Because of the wide diversity in ecology, especially in the former Soviet Union, new knowledge will be required regarding optimal tillage techniques, soil conservation measures, plant varieties, and crop rotation, as well as livestock species and their numbers and stocking rates. Socialist agriculture largely ignored ecological diversity, contributing to large-scale degradation of agricultural land.

Some caveats

Much of the existing evidence suggests that it makes good economic and environmental sense to focus, at least initially, on the "win-win" environmental possibilities that accompany the shift from a centrally planned economy to a market economy. However, this should not lead to complacency. Even if large potential environmental gains are possible from an "all-good-things-go-together" strategy, this possibility does not ensure that the shift to market-based growth policies will be environmentally sustainable. Experience in the developed countries shows that attaining desirable improvements in environmental quality requires more than getting such policies right. It also requires environmental legislation and a public-sector environmental regulatory structure that comes at a cost.

For several reasons, this lesson has been downplayed by economists and donor agencies working in the former socialist countries. Because significant improvements in environmental quality appear to be achievable without erecting a set of costly environmental institutions, there is some justification for focusing on no-cost or low-cost interventions. This predilection is reinforced by concern for weaning public-sector actors in the former communist countries from their dependence on command-and-control approaches to economic and environmental management.

But this concern is not sufficient justification for failing to invest in public-sector regulatory institutions. Failure to do so could have several severe environmental consequences. Even though the former communist countries are moving their economic policies in the direction of market-oriented growth, this process is incomplete and subject to reversal. In this situation, it may be important to build an environmental regulatory structure to make the impact of actual policies more environmentally friendly. If this is not done, we risk the worst of both worlds—less growth and more environmental degradation.

Moreover, unless existing environmental institutions become more efficient, there is some danger of long-term erosion in public and private commitments to environmental protection. Such erosion could deprive a more efficient public-sector environmental regulatory institution of its rightful place in long-term environmental management. But we should not fool ourselves that it will be easy to build such an institution. Much remains to be done. We had better use the breathing space provided by an immediate focus on capturing win-win opportunities to start now.

Michael T. Rock is associate director of the Center for Economic Policy Studies at Winrock International in Arlington, Virginia.

Milton Keynes UK
Ingram Content Group UK Ltd.
UKHW022109141024
449569UK00031B/1846